Applied Agriculture

Co-ordinating Editor:
B. Yaron

Editors:
B.L. McNeal, F. Tardieu,
H. Van Keulen, D. Van Vleck

Springer-Verlag Berlin Heidelberg GmbH

Samuel Dasberg · Dani Or

Drip Irrigation

With 61 Figures and 24 Tables

 Springer

Dr. Samuel Dasberg
Agricultural Research Organization
Institute of Soil, Water and Environmental Sciences
The Volcani Center
Bet Dagan 50250
Israel

Dr. Dani Or
Departments of Plants, Soils and Biometeorology
and Biological and Agricultural Engineering
Utah State University
Logan, UT 84322-4820
USA

ISSN 1433-7576

Library of Congress Cataloging-in-Publication Data
Dasberg, S.
     Drip irrigation / Samuel Dasberg, Dani Or.
        p.   cm. -- (Applied agriculture, ISSN 1433-7576)
     Includes bibliographical references and index.

     1. Microirrigation.   I. Or, Dani, 1955-   . II. Title. III. Series.
   S619.T74D37   1999
   631.5′87--dc21

ISBN 978-3-662-03965-6   ISBN 978-3-662-03963-2 (eBook)
DOI 10.1007/978-3-662-03963-2

© Springer-Verlag Berlin Heidelberg 1999
Originally published by Springer-Verlag Berlin Heidelberg New York in 1999.
Softcover reprint of the hardcover 1st edition 1999

Production: PRO EDIT GmbH, Heidelberg
Cover design: design & production, Heidelberg
Typesetting: Best set Typesetter Ltd., Hong Kong
SPIN 10135491   31/3137   5 4 3 2 1 0 – Printed on acid-free paper

Preface

The aim of this book is quite ambitious: here, we attempt to bridge the gap between soil physicists, agronomists, horticulturists, hydraulic engineers, designers, manufacturers and users of drip irrigation systems. We believe that progress in drip irrigation hinges on the contributions of professionals made in all related disciplines and their cooperation.

The last decade has seen great development in the field of drip irrigation, although the drip-irrigated area has not increased at the same rate as in the previous decade. However, our understanding of the processes involved in water and solute distribution and in plant response has increased vastly. The tools for optimal design of drip systems have improved tremendously. The main progress has been in the development and in the manufacture of sophisticated equipment; not only improved types of emitters and laterals, but also auxiliary equipment such as new filtration systems, controllers and sensors. In this book we highlight the need to maintain a proper balance between the hydraulic design of drip systems and aspects of their management and maintenance. Drip irrigation has a potential for high water use efficiency, but many well-designed systems suffer from bad management.

We are indebted to the late Eshel Bresler for his contribution to our understanding of water and solute movement under drip irrigation and its application to system design. Some parts of a previous publication entitled "Drip irrigation manual" authored by S. Dasberg and E. Bresler were used in the preparation of this book. Special thanks are due to Dr. Markus Tuller (Utah State University) and to Ms. Eleanor Watson (Utah State University) for their able assistance in manuscript preparation and preliminary reviews of the material. We also thank Drs. Yossi Shalhevet, Reuven Steinhardt and Gerald Stanhill (ARO, Institute of Soil, Water and Environmental Sciences) for their reviews and useful comments. The partial support of the US-Israel Binational Agricultural Research and Development Fund (BARD) through grant IS-2131-92RC provided the foundation for the collaborative effort leading to this book and is gratefully acknowledged.

Finally, we would like to thank our families for bearing with us during our preoccupation with the writing of this book.

Bet Dagan/Logan   Samuel Dasberg
March 1999   Dani Or

Contents

Introduction

1.1
Definitions

Drip irrigation is defined as the application of water through point or line sources (emitters) on or below the soil surface at a small operating pressure (20–200 kPa) and at a low discharge rate (1–30 l/h per emitter), resulting in partial wetting of the soil surface. In the literature, "trickle" is used interchangeably with "drip". In this book we exclusively use the term drip irrigation.

Another related, but more broadly defined, term is microirrigation: water is applied not only by emitters on or below the soil surface, but also by sprayers, microjets or bubblers above the soil, conveying water through the air and not directly to the soil, also resulting in partial wetting of the soil surface. Microirrigation differs from sprinkle irrigation by the fact that only part of the soil surface is wetted. We will confine ourselves to drip irrigation, although most of the statistics on drip irrigation refer to microirrigation in its broader sense.

1.2
Historical Overview

The origins of drip irrigation are manifold. Some irrigationists claim that the forerunner of drip irrigation is subsurface irrigation by means of a drainage system as practiced in Germany more than a century ago. Many horticulturists have noted that plants thrive near dripping faucets or other water leaks. One of the first references to "trickle irrigation" in the USA can be found in an early work of Reuther (1944), who noted the beneficial effect of continuous drip irrigation on date palms in the Coachella valley.

Technological development on an industrial scale came about only with the "plastic revolution" after World War II. One of the earliest developments was in commercial tomato culture in glasshouses in England between 1945–1948 (Searle 1954; Waterfield 1973). Water was supplied to individual plants by screw-type nozzles with a discharge of 1–2 l/h (Cameron Irrigation Ltd.) or through long 1-mm diameter "spaghetti" tubes providing a steady trickle (Volmatic system of Volmar Hanson in Denmark). The "trickle" system in operation in a tomato glasshouse in England some 50 years ago shown in Fig. 1.1 has all the attributes of a contemporary drip layout: screw-like nozzles (emitters) inserted in rubber tubes (laterals) and a fertilizer injection system (Searle 1954). Similar development occurred in the USA, as reported by Bucks and Davis (1986). In Israel, Blass (1971) developed a system based on the same

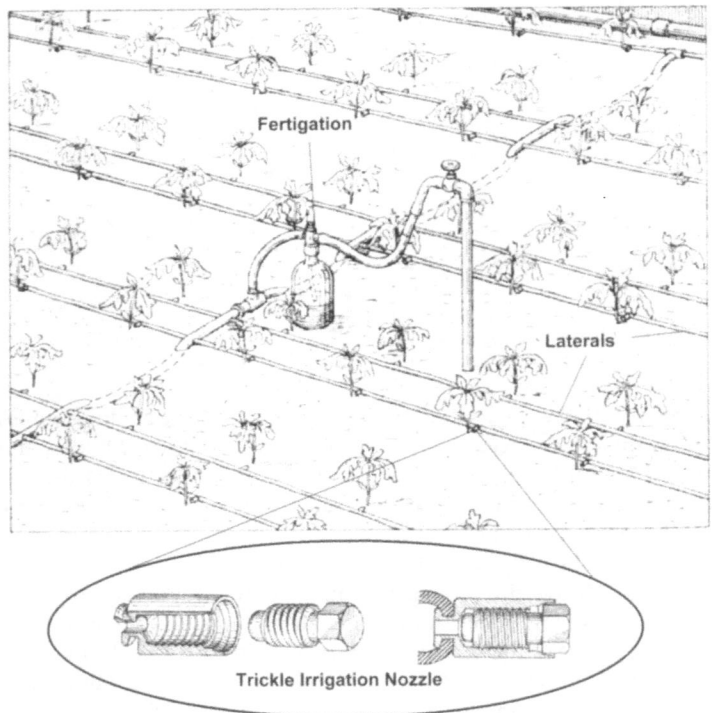

Fig. 1.1. A typical drip irrigation layout for a commercial glasshouse tomato crop (Searle 1954)

principle in the early 1960s. He initially built a subsurface long-tube system, which evolved into a small spiral-type long-path emitter. The first subsurface installation system in Israel developed clogging problems, especially by root penetration into the emitters. Zohar (1971) was the first irrigationist to successfully experiment with drip irrigation under field conditions. He placed the emitters on the soil surface, thereby avoiding the clogging problem by roots. This approach resulted in rapid development of the method. The work of Goldberg et al. (1976) promoted further progress in this field.

An account of the early history of drip irrigation in Israel is given by Blass (1971), and corresponding developments in the USA are discussed in Chapin (1971). It is noteworthy that drip irrigation was not even mentioned in the monograph of the American Society of Agronomy, *Irrigation of Agricultural Lands* (Hagan et al. 1967). Reference to the method was given in a monograph published six years later (Heller and Bresler 1973). Chapin (1971), Davis (1983) and Gustafson (1977), among others, contributed to the promotion and development of drip irrigation in the USA. In Australia, Black (1976) carried out much of the early work on drip irrigation.

Drip irrigation can be applied by individual emitters or point sources to widely spaced crops, such as orchards or vines, and sometimes with more than one emitter per plant. Row crops are frequently irrigated with line sources consisting of closely spaced emitters on a lateral row of porous pipes. Recently, there has been a revival of subsurface drip irrigation (SDI), as some of the early problems encountered with this system were overcome.

The First International Meeting on Drip Irrigation was held in Israel in 1971, at which 24 papers were presented. In 1974, 83 papers were presented at the Second International Drip Irrigation Congress in San Diego; while at the Third Congress held in Fresno, California in 1985, 160 papers were presented. Dasberg and Bresler (1985) wrote in the preface of the Drip Irrigation Manual, which was a forerunner of this book: "Drip irrigation has come of age. After the initial excitement and high hopes raised by the pioneers of the system in the early seventies, the time has come for a balanced assessment of the merits and potentials of drip irrigation." Another textbook, entitled *Trickle Irrigation for Crop Production* was published shortly thereafter (Nakayama and Bucks 1986). Since then, an International Microirrigation Congress has been held in Adbury, Australia in 1988 with 89 presentations, and, more recently, in 1995, in Orlando, Florida, where 156 papers and posters were presented.

Areas under drip irrigation expanded rapidly during the 1970s; whereas only 600 ha were drip irrigated in the USA in 1970, the area had increased to 75 000 ha by 1976 (Gustafson 1977). Table 1.1 gives the area under drip irrigation in 1982

Table 1.1. Total irrigated area and area under drip irrigation in 1982 and in 1992 (Bucks 1995)

Country or State	Total (in 1000 ha) 1982	1992	Drip (in 1000 ha) 1982	1992	Drip (% of total) 1982	1992
California	409	3 629	105	99	2.6	5.5
Florida	944	823	19	182	3.4	5.0
Hawaii	71	59	12	45	17.1	90
Georgia	469	497	0	24	0	4.8
Texas	3 200	2 471	10	24	0.3	1.0
Arizona	466	386	2	12	0.4	3.1
Michigan	171	214	15	22	8.8	10.3
Washington	762	828	0	12	0	1.5
Total USA	24 811	23 640	185	606	0.7	2.6
Australia	1 500	2 069	20	147	1.3	7.1
S. Africa	1 020	1 128	44	144	4.3	12.8
Israel	203	180	82	104	40.0	57.8
Spain	3 100	3 403	0	160	0	4.7
Italy	2 900	3 150	10	79	0.3	2.5
Egypt	2 540	2 645	0	68	0	2.5
Mexico	4 800	6 100	2	60	0	1.0
Japan	3 010	2 802	0	57	0	2.0
Portugal	630	633	2	24	0	3.2
India	35 150	45 800	22	71	0	0.2
France	960	1 190	22	51	2.3	4.3
Thailand	3 320	4 400	0	41	0	0.9
Chile	1 257	1 268	0	26	0	2.0
Cyprus	30	37	6	25	20	67.6
China	44 594	49 030	8	19	0	0
Jordan	38	65	1	12	2.6	18.5
Morocco	1 230	1 280	4	10	0	0.8
Other			50	100		
Total	131 216	148 830	457	1 861	0.3	1.3

and in 1992 based on the statistics given by Bucks (1995), plus additional information from other countries, as percentages of the total irrigated area, according to the *FAO Production Yearbook* (1982, 1992). These data show that in 1982 only 0.3% of the total irrigated area in the world was drip irrigated. In spite of the fact that the worldwide area under drip irrigation increased more than fourfold to almost 2 million ha during this decade, the percentage of total irrigated area was still only 1.3%. Bucks (1995) predicted that by the year 2000 the drip-irrigated area in the world will reach 3 million ha, representing 2% of the total irrigated area in the world.

Drip irrigation occupies a significant portion of irrigated areas only in specific locations and for special crops, such as sugar-cane fields in Hawaii, glasshouses in the UK, high value vegetable field crops, cotton fields in Israel and tree crops in many parts of the world. The reason for this limitation in the development of the method is economical: the high cost of equipment, its installation and maintenance, in conjunction with the low price of water in many parts of the world. Only when water is severely limited, highly priced and when its distribution is tightly controlled by a central agency, is the high degree of control of water application offered by drip irrigation of real economic advantage.

1.3
Specific Attributes of Drip Irrigation

1.3.1
Advantages

In attempting to assess the specific advantages of drip irrigation and its potential, one is confronted with the problem of choosing a common basis for comparison. In the traditional irrigation systems (surface or sprinkle), water is applied at relatively large intervals, so that evapotranspiration and water extraction by plant roots between irrigations lead to the development of soil water deficit. With solid-set and center-pivot systems, water can be applied in small quantities at high frequencies. In addition, with drip or solid-set minisprinklers, irrigation is applied to only part of the surface area. These differences in soil water regime and application geometry make an objective comparison of the systems very difficult. Nevertheless, we will use the following four criteria as a basis of comparison: (1) the total quantity of water applied, (2) the amount applied per unit wetted area, (3) the plant water use (disregarding frequency of application or surface wetting pattern), and (4) yield response. Increased yields obtained with drip irrigation (as reported by many authors) can be attributed to several factors: higher water use efficiency because of precise application directly to the root zone and lower water losses due to reduced evaporation, runoff and deep percolation; less fluctuations in soil water content resulting in the avoidance of water stress; improved cultural practices such as fertilizer application, weed control and others. We will try to enumerate some of the specific advantages of the drip system as compared to sprinkler and other irrigation systems, bearing in mind the limitations of such comparisons.

1.3.1.1
Controlled Application

The high degree of water application control is the main advantage drip irrigation offers. The system is comprised of a very large number of water sources per unit area with high uniformity of discharge. Drip irrigation is usually applied with a solid-set (stationary) system. This ensures accurate and localized application, at predetermined amounts and at constant rates. The time of day chosen for applying water by drip is not limited by wind speed as in the case of sprinkler irrigation, and little uniformity is lost due to runoff. However, water distribution around each emitter is not uniform, as will be shown later. It is possible to compare field application efficiencies of surface, sprinkler and drip irrigation using the definition according to Bos (1979), namely the ratio between water at the field inlet and water needed to maintain the soil water level above a minimum required for the crop. Wu and Gitlin (1975) concluded that an application efficiency of 90% could easily be achieved for drip irrigation as compared to 60–80% for sprinkler and 50–60% for surface irrigation. This calculation assumes that variation in the emitter flow does not exceed 20%, which is a very conservative criterion according to Solomon and Keller (1978). The manufacturing variation of many of the modern emitters has coefficients of variation less than 5%, resulting in a uniformity coefficient of more than 96%.

1.3.1.2
Maintenance of High Soil Water Potential in the Root Zone

Irrigation should supply water at a rate sufficient to satisfy crop evaporative demand by maintaining high matric and osmotic potentials of the soil water, which minimizes water and osmotic stresses. In order to achieve this goal of high matric potential, the water content should be kept as high as possible without causing soil aeration problems. In the case of surface or sprinkle irrigation, a certain depth of the soil may be saturated during irrigation, resulting in problems of aeration. During drip irrigation, part of the root zone is below field capacity, providing an adequate oxygen supply to plant roots. Furthermore, frequent irrigation will cause frequent leaching of excess salts to the periphery of the root zone, maintaining a high osmotic potential. High frequency irrigation (Rawlins and Raats 1975) is one of the main characteristics of drip irrigation. In traditional low-frequency irrigation systems, such as furrow, flood or portable sprinklers, the fluctuations in soil matric and solute potentials are relatively large. Frequent water application with these systems is not possible for several reasons: the need for a minimum flow to cover the whole area (making low-volume water applications impractical), water rights or stream flow making scheduling necessary, and the fixed labor costs inherent in each water application (economic constraints). The permanent (solid set) systems with their low water pressure and low flow rate, such as drip and minisprinklers, permit high frequency irrigation resulting in a continuously high soil water content and low solute concentration (high total soil water potential) and in possible benefits to plant growth (Rawitz 1970; Bresler 1977).

1.3.1.3
Partial Soil Wetting

With drip irrigation as applied in orchards or row crops, only a portion of the soil surface is wetted. The fraction of root volume wetted may be somewhat larger, due to lateral movement of water below the soil surface. This enables root aeration at the fringes of the wetted volume. Examples in the literature describe trees able to thrive on a very small root volume, provided adequate water and nutrients are supplied. Black and West (1974) showed that 25% of the root system of a young apple tree could absorb 75% of the water taken up by a tree with an unrestricted root system. Willoughby and Cockroft (1974) reported similar findings for mature peach trees. They observed the proliferation of new roots in the vicinity of the emitters, a fact observed by many drip irrigators. Many examples in the literature show the high efficiency of fertilizer uptake when applied through the drip system to a restricted soil volume (Frith and Nichols 1974).

Because of partial soil wetting, less water is lost by direct evaporation from the soil surface, which is one of the specific advantages of drip irrigation. To estimate the extent of this water loss, a simplified model was suggested by Ritchie (1972) and extended by Tanner and Jury (1976), using the energy-balance approach. During the first drying stage, when evaporation depends only on the evaporative demand, soil evaporation E_s may be described by

$$E_s = \alpha_E E e^{-\beta \text{LAI}}, \tag{1.1}$$

where E is the evapotranspiration according to Penman without advection; α_E is the wet bare soil albedo, equal to 1.0 for full cover vegetation; β is an empirical factor, usually equal to 0.4; and LAI is the leaf area index (leaf area to soil surface area ratio).

At the second stage of drying, when evaporation is limited by soil hydraulic conductivity, evaporation from the soil surface can be calculated from

$$E_c = C(t - t_c)^{1/2}, \tag{1.2}$$

where E_c is the cumulative soil evaporation; C is a soil constant which varies with season; and t_c is the time at which soil water content reaches a critical value and the stage of decreasing rate of evaporation starts. During the initial stage of crop development, when LAI < 1, soil evaporation calculated from this model may reach more than 70% of the total water loss. When full cover is reached, however, direct evaporation from a fully wetted soil is generally less than 10% of the total loss. The evaporation from a fully wetted row crop field for the whole growing season would be about 30% of total evapotranspiration. Soil surface evaporation for drip irrigated corn was measured recently by comparing irrigation with and without soil cover with the help of Time Domain Reflectometry, TDR (Coelho and Or 1996). The results showed surface evaporation to be only 7–10% of the total water loss (ET) with full crop cover. With subsurface drip irrigation (SDI), soil evaporation losses should be minimal. The ratio of bare soil evaporation to evapotranspiration of a grass canopy as measured by weighing lysimeters was 0.06 for SDI, compared with 0.18 for drip irrigation applied to the soil surface (Phene et al. 1989). The upper bound of evaporation from a buried source was calculated by Philip (1991a). He found that the relation between total evaporation from unit area E and emitter flow Q in l/h, $E/Q = e^{-z\alpha}$, where z is the emitter depth and α is a soil characteristic ranging from 0.2 for sand to 0.02 for clay (see Table 3.1). For clay-loam soil, with irrigation of 3 hours per day, this would result in $E = 0.07Q$, similar to the experimental results of Phene et al. (1989).

Several empirical expressions have been proposed for relating total evapotranspiration E_t to pan evaporation E_{pan} (Aljibury et al. 1974; Shearer et al. 1975; Walker et al. 1976) according to

$$E_t = \delta A E_{pan}, \tag{1.3}$$

where A is the percentage shaded area in the orchard and δ is an empirical constant (generally equal to 1).

Inserting the relevant values in Eqs. (1.1) through (1.3), partial wetting of the soil surface with drip irrigation can be shown to result in a 20–40% less water use through a reduction in the direct soil evaporation in non-wetted areas of row crops and orchards. However, it should be noted that with drip irrigation the water is usually applied at a higher frequency than with sprinkler or furrow irrigation. A careful comparison of daily drip and 10-day interval furrow irrigation of canning tomatoes using large weighing lysimeters (Pruitt et al. 1981) showed no substantial difference in total seasonal ET between the irrigation methods. The daily ET from the furrow-irrigated plots was 30% higher than from the drip irrigated plots during the first 2 days after each furrow irrigation, which implies higher soil surface evaporation. During the 8 subsequent days, ET losses under daily drip irrigation exceeded those for furrow irrigation by 8–9% (the soil surface was continuously wet). Thus, partial surface wetting of row crops by daily drip irrigation does not necessarily result in any seasonal water saving.

The management advantages of partial wetting of the soil are as follows:

1. Restriction of weed growth to the wetted area thereby reducing the cost of weed control and competition for water and nutrients by weeds.
2. Unrestricted travel in the permanent dry strip between the rows in the orchard allows spraying, picking and harvesting to be carried out with minimum soil structure damage due to puddling and compaction of wet soils.

1.3.1.4
Maintaining Dry Foliage

Dry foliage retards the incubation and development of many plant pathogens (Yarwood 1978). Therefore, less frequent pesticide and fungicide application is required under drip irrigation, and the chemicals are not washed from the leaves by irrigation water. An additional advantage of dry foliage is the avoidance of leaf burn when irrigating with saline water (Bernstein and Francois 1975) and the possibility of using reclaimed sewage water without leaf and fruit contamination. Moreover, dry foliage eliminates direct evaporation loss of water from the canopy, although such evaporation is sometimes desirable for microclimate modification. Shalhevet et al. (1983) showed a 10% reduction in water loss due to avoidance of canopy wetting of potato by drip compared with sprinkler irrigation.

1.3.1.5
Use of Low-Quality Water

Several carefully conducted experiments have been carried out comparing drip, sprinkler and furrow irrigation using water of different salinity. Bernstein and Francois (1973) varied the amounts of water applied according to plant needs; they obtained similar yields of peppers for three irrigation methods when using low salinity water but obtained a yield reduction of 13% for furrow and 59% for sprinkler, compared with drip when applying brackish water. Goldberg et al. (1976) reported increases in yields of several vegetable crops under drip, compared with sprinkler and furrow irrigation, with saline water having an electrical conductivity of 3–4.5 dS/m (mmho/cm). The yield of sorghum irrigated by drip irrigation with water containing 1600 mg/l salts was significantly higher than when the same water was applied by surface irrigation (Seifert et al. 1975). The following factors contribute to the good results obtained with saline water using drip irrigation:

1. Foliar absorption of salts and leaf burn are avoided, as discussed above.
2. The increase in salt concentration of the soil solution resulting from soil drying between two successive irrigations is less in high frequency drip irrigation than in traditional low-frequency irrigation systems.
3. Salts from the wetted section are continuously leached from the active root zone and accumulated at the periphery of this zone (Yaron et al. 1973; Bresler 1975 Hoffman et al. 1980).

The degree of permissible water salinity for use in drip irrigation depends on water quantity, soil hydraulic properties and crop tolerance. Shmueli (1975) noted reduced growth and yields for peas, sweet corn and tomatoes irrigated by drip with water having electrical conductivity of 3.5 and 6.4 dS/m, respectively; whereas peppers showed no yield reduction. Fruit size was affected by the saline water at all levels with all the tested crops.

Another advantage of drip irrigation is the possibility of utilizing sewage water after secondary treatment and adequate filtration (Oron et al. 1979; Bielorai et al. 1980). Drip irrigation reduces the hazard of aerosols, which might deposit disease-causing microorganisms on the foliage (Shuval 1977: Katzenelson and Teltch 1976). The risk of deep percolation of soil contaminants is reduced because of the high water application efficiency resulting in minimum drainage. The lack of direct contact between the wastewater and fruits or canopy is of great importance (Feigin et al. 1991; Oster 1994).

1.3.1.6
Economic and Energy Benefits

The cost of installation of a permanent drip system in a row crop is usually higher than for a sprinkler system, even solid-set, mainly because the laterals are more closely spaced (Sefarim and Shmueli 1975; Reed et al. 1977). Letey et al. (1990) carried out an economic analysis of several irrigation systems for cotton in California. The annual costs for pressurized systems (subsurface drip, LEPA and sprinkling) were higher than for surface irrigation. If drainage disposal costs were taken into account, the profitability of drip irrigation became high because of the high irrigation uniformity. Energy requirements (pumping cost) and labor cost are generally lower for drip systems. In orchards or other widely spaced crops, the cost of installing a drip system may be lower than for a sprinkler system, since, because of the lower discharge of emitters, smaller diameter laterals can be used (Karmeli and Keller 1975). Since it is difficult to compare costs in these times of rapid price changes, it is feasible to compare drip with other irrigation systems on an energy-expenditure basis, as was done by Batty et al. (1975). In Table 1.2, we provide some data from their analysis, based on several different irrigation systems designed for a 64-ha farm. Inputs into the systems were expressed in terms of energy units per unit area, MJ/ha. The data in Table 1.2 show the advantage of drip over sprinkler irrigation with regard to pumping energy requirements. The difference in this energy value for the two systems is more pronounced if the higher water application efficiency of drip irrigation is taken into account. The energy expenditure for labor is negligible compared with pumping and installation energies. Stibbe (1986) compared the energy requirement of drip irrigation with sprinkling for cotton growth in Israel. He found that the annual amortization of energy invested in a drip system was 13.4 GJ ha^{-1} y^{-1}, compared with 6.6 for sprinkling. Pumping energy for drip was lower by 8–10 GJ ha^{-1} y^{-1}, compensating for the high energy investment in the drip system (Stibbe 1986).

Labor costs, however, are the major expenditures in furrow and portable sprinkler systems, while they are less than 10% of the total costs in the permanent drip or solid-set. A comparison (Funt et al. 1980) between drip and travelling gun sprinkler irrigation for supplemental irrigation of orchards under humid conditions showed that total water use, installation and pumping costs for the drip system were less than half of those for the sprinkler system. Labor costs were similar in both cases.

A more recent economic analysis of subsurface drip irrigation (SDI) and center pivot (CP) for field corn was carried out in Kansas (Dhuyvetter et al. 1995). The system cost for SDI was $1405/ha compared with $951/ha for CP. Assuming equal corn yields for both systems (11 Mg/ha), the CP system had an annual economic advantage of $56 over the SDI system. The SDI system irrigated more area (unirrigated corners in the CP system) and generated more returns on the investment than the CP system; however, it could not overcome the greater costs associated with the higher initial investment. Annual return on both systems is equal only when crop yield under SDI is 0.6 Mg/ha higher than the CP yield.

Table 1.2. Total energy inputs in MJ/ha for five irrigation systems, assuming 915 mm water requirement and zero lift (Batty et al. 1975)

Irrigation system	Installation	Pumping	Labor	Total
Furrow	1858	498	3.9	2361
Permanent sprinkler	5102	7958	0.8	13060
Hand-moved sprinkler	1649	8309	5.0	10008
Center-pivot sprinkler	4014	8929	0.8	12943
Drip	5493	4839	0.8	10323

Finally when corn yields exceed 14 Mg/ha, the SDI system becomes more economical.

Some operational advantages of drip over other pressurized systems are:

1. The possibility of irrigating any time during the 24-h day regardless of wind velocities
2. Lower pressure requirements
3. Lower water flow per unit area than for sprinkler, requiring smaller diameter of mains and laterals.

These advantages can be translated into substantial economic benefits in the design of drip systems compared with sprinkling.

1.3.1.7
Fertilizer, Herbicide and Pesticide Application (Fertigation and Chemigation)

With drip irrigation, it is possible to apply fertilizers in solution along with the irrigation water, a practice known as fertigation. This process has several advantages over the traditional methods of broadcasting and mechanical incorporation:

1. The method is labor and cost saving.
2. The application is more precise, being restricted to the wetted area where the active roots are concentrated; this leads to more efficient utilization and minimum waste of fertilizers by leaching and subsequent pollution problems.
3. The concentrations and amounts of individual nutrients can be more readily adapted to plant needs according to stage of development and climatic conditions.

However, fertilizers must be completely soluble in water in order to be distributed evenly through the drip system (Grobbelaar and Lourens 1974). Chemicals of low solubility may precipitate causing blockage of the emitters. There is usually no solubility problem with nitrogen and potassium compounds (Miller et al. 1975). Phosphorus is usually added in soluble forms as potassium orthophosphate, as mono ammonium polyphosphate or as diammonium phosphate (Rauschkolb et al. 1976). Microelements may be added in chelate form.

An experiment with labeled nitrogen demonstated that the uptake of fertilizer N by tomatoes was more efficient when nitrogen was applied through the drip system than when applied by furrow irrigation. On the other hand, with no fertilizer application, the yields were significantly lower with drip than with furrow irrigation (Miller et al. 1981). This was possibly a result of root proliferation in a limited soil volume adjacent to the point source where N was rapidly depleted, necessitating a continuous supply of nutrients with the irrigation water. Some instances of drip irrigation failure may have been due to this factor being ignored.

The drip irrigation system is well suited to the application of herbicides and for soil-born diseases and pests, since localized application only in the wetted area results in the chemicals being more effective at lower concentrations (Gerstl et al. 1981).

1.3.1.8
Adaptation to Marginal Soils

Drip irrigation has been used to irrigate marginal soils and terrain that could not have been irrigated by other methods (Marsh 1977). For example, avocados are being grown successfully in San Diego County on steep, rocky hills under drip irrigation. Soil with high permeability and low water-holding capacity, such as sands, desert pavement and leached tropical soils, adapt poorly to surface or sprinkler irrigation, but may be irrigated successfully with drip systems. Drip irrigation has been proven to be an efficient and effective technique for establishing vegetation on steep slopes of abandoned mines, road embankments, etc., without erosion hazard (Bengson 1977). It is also very suitable for irrigating slowly permeable soil (Grimes et al. 1976) and irregular small plots.

1.3.1.9
Adaptation to Landscape Irrigation

Small, irregularly shaped and narrow lawn and landscaped areas are difficult to irrigate by sprinkling resulting in overspray of paved surfaces and lack of uniformity. Drip irrigation enables water to be applied with high uniformity and may eliminate runoff and overspray. Subsurface drip on turfgrass and sportfields does not interfere with the continuous use of the area. (Rochester 1995; Zoldoske et al. 1995).

1.3.1.10
Adaptation to Protected Crops

Drip irrigation was first commercially used for glasshouse culture in England (Waterfield 1973). It provides controlled application of water and nutrients for individual plants without foliage wetting, which is an important feature for high-value crops such as flowers, potted plants and glasshouse vegetables. A high proportion of these crops are drip irrigated in Europe and Israel. Drip irrigation is also suitable for vegetables grown on plastic mulches and under tunnels, such as strawberries and early season melons and watermelons.

1.3.2
Limitations

1.3.2.1
Clogging of Emitters

The main problem associated with drip irrigation is clogging of the emitters. Emitters usually have orifice diameters of only 0.5–1 mm and are thus vulnerable to clogging by root penetration, sand, rust, micro-organisms or other impurities in the irrigation water or by the formation of chemical precipitates (Ford 1977). In wastewater irrigation, the main causes of clogging are suspended matter, chemical precipitation and bacterial growth (Adin and Sacks 1991), or protozoa (Ravina et al.

1992; Sagi et al. 1995). The best way to reduce or prevent clogging is by adequate filtration. The different filtering systems and the problems associated with them will be discussed in Section 2.4.

Root penetration into emitters occurs only in buried drip systems. It can be prevented by the injection of treflane, a herbicide strongly adsorbed by the soil colloids which prevents root proliferation in the near vicinity of the emitter (Bucks et al. 1981; Phene et al. 1983). Recently, treflane-impregnated emitters that slowly release the herbicide were introduced to the market (Geoflow Inc. California). It has been shown that root intrusion into buried emitters is not a problem if the system is operated at high frequency, causing a permanently saturated soil zone in the immediate vicinity of the emitters which roots tend to avoid (Phene 1995).

Sand from well water can be removed effectively by centrifugal separators. Suspended organic matter and clay particles can be separated with gravel filters, disk and screen filters, which have to be cleaned periodically. Filtration is not sufficient when wastewater is used for drip irrigation – chlorinating or some other method of disinfecting is needed to prevent growth of bacterial slimes and algae (Hills and Tajirshy 1995; Ravina et al. 1995). Very severe clogging problems are caused by iron and sulfur deposits (Ford and Tucker 1974) with or without growth of algae and bacterial slime. Chlorinating may solve some of these problems (McElhoe and Hilton 1974). The formation of phosphate or carbonate precipitates from bicarbonate present in the water may be prevented by pH adjustment (Gilbert et al. 1979). Some of the precipitates can be dissolved by injecting dilute hydrochloric acid into the system (Pelleg et al. 1974). Bacterial slime may be dissolved by hypochlorite injection (Nakayama et al. 1977).

Bucks et al. (1982) proposed a classification of irrigation water quality for potential clogging hazard (see Table 1.3). They classified the clogging problems into three categories: *physical* – caused by sand grains, sediment or foreign materials such as pieces of plastic or insect fragments; *chemical* – precipitation of carbonates at high pH, iron and manganese complexing with the aid of bacteria, and sulfur deposit-

Table 1.3. Proposed criteria for classifying water used in drip irrigation (Based on Bucks et al. 1982)

Factor	Slight	Clogging hazard Moderate	Severe
Physical			
Suspended solids[a]	<50	50–100	>100
Chemical			
PH	<7.0	7.0–8.0	>8.0
Dissolved solids[a]	<500	500–2000	>2000
Manganese[a]	<0.1	0.1–1.5	>1.5
Iron[a]	<0.1	0.1–1.5	>1.5
Calcium and Magnesium[a]	<20	20–50	>50
Hydrogen sulfide[a]	<0.5	0.5–2.0	>1.5
Biological			
Bacterial population[b]	<10000	10000–50000	>50000

[a] Concentration in mg/l
[b] Number of bacteria per ml

ing from H_2O due to certain bacteria; and *biological* – microbial slime, algae or plant roots. Careful determination of the causes of clogging with eight types of emitters and four water treatments over a 4-year period showed that in most cases the initial cause of flow reduction was a physical factor, with subsequent development of biological or chemical precipitates (Gilbert et al. 1981). Emitter types differ greatly as to their susceptibility to clogging depending on orifice size, path length, flow velocity, flushing characteristics and pressure compensation (Gilbert et al. 1981; Adin and Sacks 1991; Hills and Tarjishy 1995). Hills and El-Ababy (1990) evaluated the clogging characteristics of several self-cleaning emitters. These have expandable orifices allowing particles to pass at either high or low pressures. They found that all tested emitters were relatively successful in self-cleaning when the water impurities were inorganic. Organic impurities, however, resulted in gradual clogging (Hills and El-Ababy 1990). The emitter internal design can be improved by shortening, widening, rounding the edges, and removing dead areas in the flow path, plus enlarging the orifice entrance to act as a filter for larger particles (Adin and Sacks 1991).

In conclusion: the solution of clogging problems is to avoid flow reduction by preventing foreign material from entering the system, by adequate filtration and by chemical treatment of the water according to the water quality.

1.3.2.2
Salt Accumulation in Soil

When saline or brackish water is used in drip irrigation, salts tend to accumulate at the wetting front (Yaron et al. 1973; Tscheschke et al. 1974; Bresler 1975). These salts may cause serious problems to subsequent crops if irrigated by a method other than drip or in arid areas where rainfall is insufficient to leach the accumulated salts. The problem can be reduced by leaching of the accumulated salts by sprinkler or flood irrigation, or, alternatively, if the subsequent crop is to be drip irrigated, by returning with the drip lateral to the same location (Goldberg and Uzrad 1976). Salinity may be a problem with subsurface drip irrigation. Above the drip line, no leaching occurs during irrigation, and salts accumulate during the irrigation season. Moderate rainfall may move salts into the root volume. Only high rainfall could move the salts below the root zone (Hanson 1995).

1.3.2.3
Mechanical Damage

Damage to the drip system is sometimes caused by man (implements or vandalism) or by animals (birds or mammals making holes in the laterals while searching for water). This damage may be partially prevented by covering the laterals and emitters with a shallow layer of soil, but by doing so, the problems of clogging by roots may occur and, furthermore, the performance of the emitters cannot be easily observed. Trapping, or repelling, the fauna or providing them with an alternative source of water is sometimes effective.

Mechanical damage may also be induced by annual removal and subsequent installation of the laterals, by tillage implements or by thermal expansion and contraction, which may disconnect in-line emitters.

1.3.2.4
Lack of Microclimate Control

Sprinkler systems may provide some microclimate control under adverse conditions. Sprinkling can be an effective measure in frost protection under conditions of radiation frost (Landers and Witte 1967). It has been shown that wetting of the canopy in a greenhouse may decrease midday water stress and transpiration rate (Plaut and Zieslin 1977). The latter effect is only minor in the open. The lack of foliage wetting in drip irrigation is a disadvantage under extreme environmental conditions.

1.3.2.5
Irrigation for Seed Emergence

Drip irrigation wets only part of the soil surface. This may pose a problem for the first irrigation of vegetable or field crops, especially with subsurface drip irrigation. Complete wetting of the soil is necessary in order to enable germination of the seeds. This can be achieved by applying large amounts of water, causing the wetted volumes of the individual emitters to overlap, or by moving the surface laterals during irrigation for this purpose. In most cases, especially in sandy soils, emergence irrigation with a drip system is impractical and, in order to achieve good germination, water should be applied by sprinkling or flooding. Additional labor and costs are incurred in this operation.

1.3.2.6
Operational Constraints

High technical skills are required for the proper design and maintenance of drip irrigation systems. The filtration requirements are stringent and should be designed according to fluctuations in water quality. Careful monitoring of the filtration system, the operating pressures and the emitter flow rates is required. Drip systems have a limited buffering capacity because of the limited wetted root volume. Therefore, any malfunction of pumping, filtration, fertilizing or chlorination equipment or any leak in mains or laterals can have a disastrous consequence, if not corrected in a timely manner. This is especially true for a subsurface system, where the emitters are buried and any failures caused by clogging are difficult to observe and still more difficult to repair.

Drip System Components

2.1
Emitters

The emitter is a device used to dissipate pressure and to discharge water at a constant rate at many points along a lateral. It is the main component of the drip irrigation system and determines its characteristics. Many types of emitters exist on the market, each with its specific properties, and may be classified according to the following criteria (Karmeli 1977): flow rate; form of pressure dissipation; and details of construction and incorporation in the lateral. Some of the older types are shown in Fig. 2.1.

2.1.1
Flow Rate and its Variation

Each emitter has a certain design flow rate, characterized by its mean at normal operating pressure and the coefficient of manufacturing variation, CV (the ratio of standard deviation to mean, Solomon 1979). The coefficient may vary from 0.02 for spiral long-path emitters to 0.40 for a porous pipe, and has a critical effect on the irrigation efficiency of the system. The flow rate is affected by pressure and temperature, as will be shown later, and obviously by clogging.

2.1.2
Form of Pressure Dissipation

The operation pressure of most emitters is in the range of 0.1–0.2 MPa. This pressure is dissipated in the emitter pathways, and reaches atmospheric pressure at the outlet in a way characteristic for each emitter, by directing the water flow through small openings. The smaller the openings, the lower the flow rates, but greater are the dangers of clogging. Several methods of pressure dissipation are employed in emitter manufacture in order to overcome the opposing constraints imposed by energy dissipation and clogging.

In long-path emitters the pressure is dissipated by flow through a long narrow path. The oldest and simplest form is the microtube or spaghetti tube (Fig. 2.1). The longer the tube, the stronger the resistance to flow. In this way a low discharge can be obtained via a tube with a relatively large diameter, and thus less prone to clogging. The functional relationships between the rate of discharge, length of path, diameter of tube and the inlet pressure will be given in the section on flow regime (2.1.5). Diameters of microtubes range between 0.6 and 1.0 mm. The energy loss (in terms of energy per

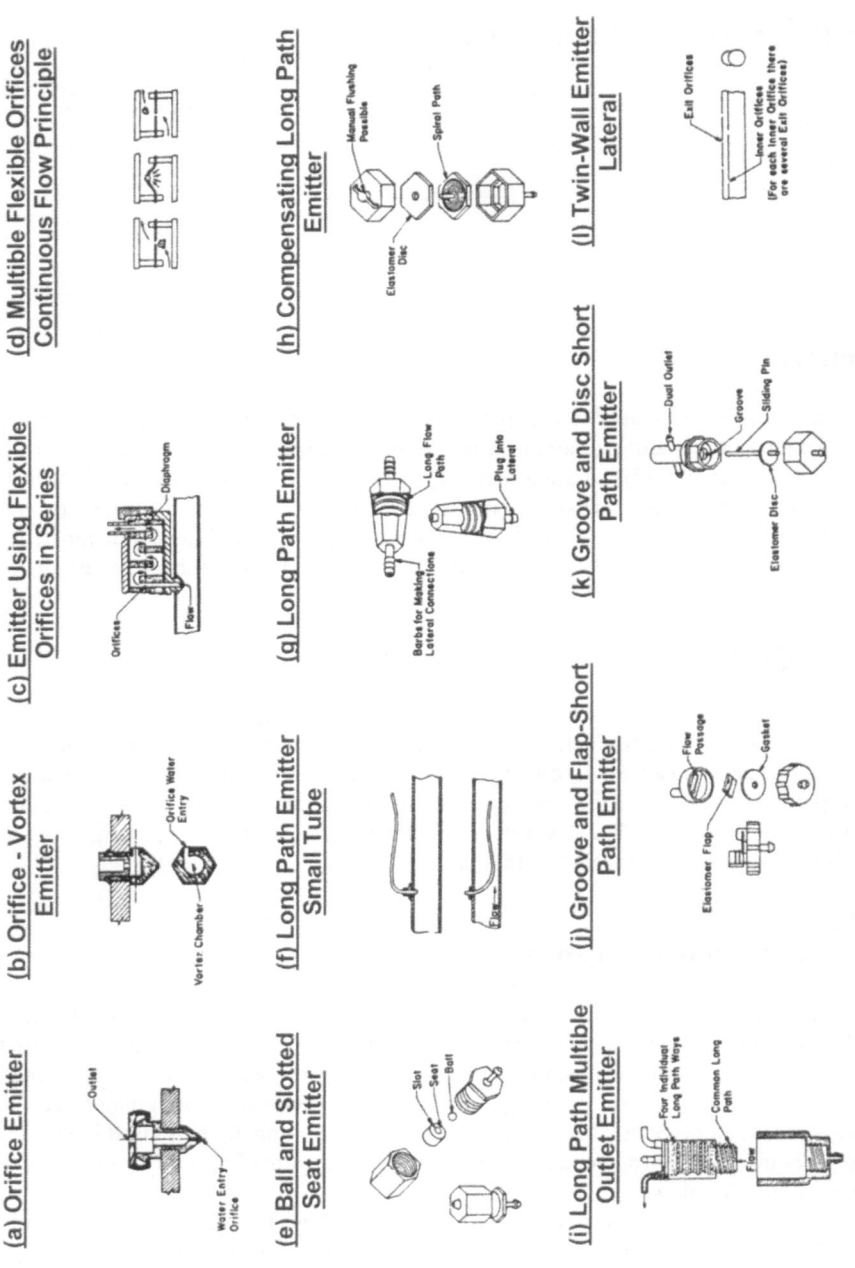

Fig. 2.1. Different emitter types (Solomon 1977)

unit weight or pressure head) in long-path emitters can be increased by roughening the flow path or by using tortuous long flow paths in the form of a spiral (screw thread) or labyrinth (Fig. 2.2). The labyrinth may be constructed in such a way as to create turbulent flow, reducing clogging potential.

In nozzle or orifice emitters, the pressure is dissipated through a small hole 0.4–0.6 mm in diameter. These emitters are very prone to clogging. The orifice diameter may be increased by creating a vortex in the flow pattern, which increases the pressure loss (Fig. 2.1). Sometimes the energy of the flowing water is dissipated through perforated tubes (small perforations in the lateral) without any emitter devices. This system is relatively simple, but has high variations in flow rates and may easily clog. Bi-wall or twin-bore (also called drip tape) is a more sophisticated form of perforated tube. This system consists of an inner tube or supply chamber operating at relatively high pressure and fitted with large, widely spaced orifices. For each inner orifice there are several outer orifices in the emission chamber for water distribution. They are manufactured with different distances between orifices (10–60 cm) and different wall thickness (0.1–0.4 mm) for shorter or longer runs (Fig. 2.3).

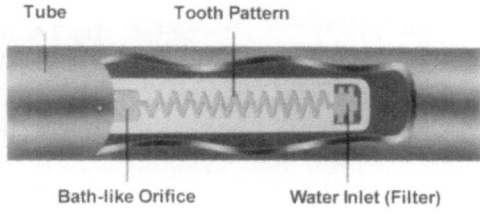

Fig. 2.2. Integral labyrinth emitter (By courtesy of NETAFIM Irrigation Equipment and Drip Systems)

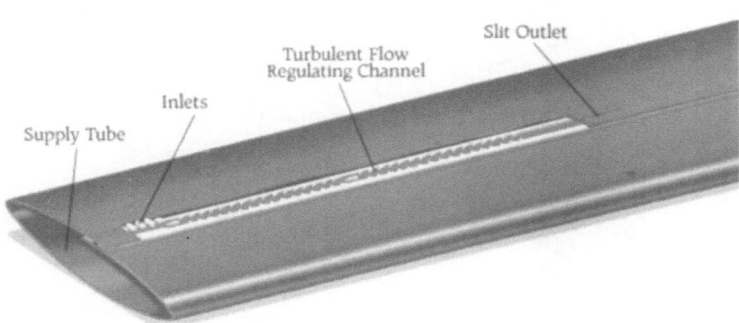

Fig. 2.3. Bi-wall driptape (By courtesy of T-SYSTEMS International)

2.1.3
Emitter – Lateral Assembly

In-line emitters are part of the lateral (see Fig. 2.4) and were the first commercially produced emitters. *On-line* emitters (Fig. 2.5), on the other hand, are inserted into the laterals. They allow more flexibility; emitters can be added if necessary (e.g. in orchards, according to tree development). They may be fitted on risers with buried laterals or may have multiple outlets for the irrigation of individual plants. Finally, *integral* emitters are molded into the wall of the laterals during the extrusion process (see Fig. 2.2). These are available in different spacings and with different wall thicknesses.

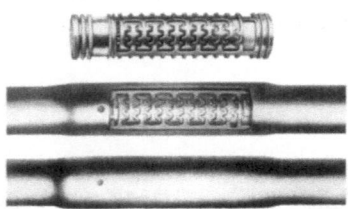

Fig. 2.4. In-line emitter (By courtesy of DRIP IN Irrigation Company)

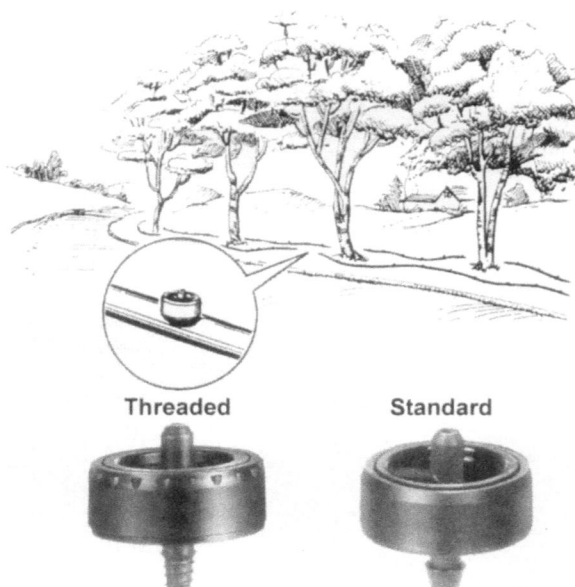

Threaded   Standard

Fig. 2.5. On-line emitters (By courtesy of PLASTRO International)

2.1.4
Discharge Regulation by Pressure Compensation

Some emitters are equipped with a specially constructed membrane or diaphragm that ensures constant discharge over a wide range of pressures (see Fig. 2.6). These emitters are usually more expensive, but allow very long laterals (more than 500 m for emitters of 1.6 l/h at 1 m distance between emitters, approximately twice as long as for non-compensated emitters). This feature enables precise irrigation of areas with varying elevation and with difficult topography.

2.1.5
Flow Regime

The flow regime (laminar or turbulent) is characterized by the non-dimensional Reynolds number Re, which for a cylindrical flow path is given by

$$Re = \frac{4Q}{\pi d \eta}, \qquad (2.1)$$

where Q is the emitter discharge rate, η the kinematic viscosity of water and d the diameter of the microtube or orifice. The three main flow regimes are usually defined as laminar, turbulent and the unstable-flow regime between them.

1. Laminar Flow Regime (Re < 2000). Microtube emitters are in this category. Their discharge or flow rate Q for any given diameter and length of flow path and inlet pressure head is given by the Hagen-Poisseuille equation

$$Q = \frac{\pi g d^4 H}{128 \eta \ell}, \qquad (2.2)$$

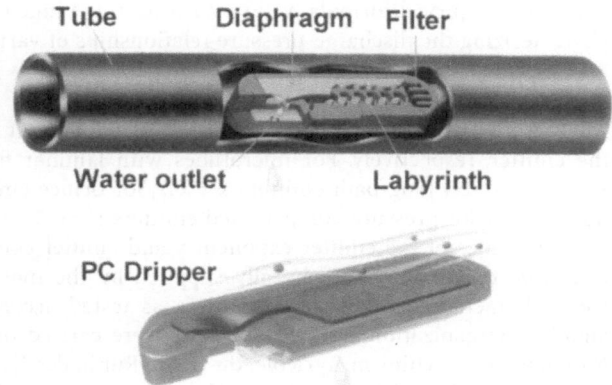

Fig. 2.6. Pressure compensated integral dripper (By courtesy of NETAFIM Irrigation Equipment and Drip Systems)

where g is the acceleration due to gravity, H is the hydraulic head at the emitter inlet (at the outlet, H = O) and ℓ is the length of the microtube. This formula shows that the discharge rate Q is inversely proportional to the length of the tube, proportional to the pressure head H for given values of d and ℓ, and proportional to d^4 for given H and ℓ. Combining Eqs. (2.1) and (2.2) with the Darcy-Weisbach formula yields

$$H = f \frac{8\ell Q^2}{\pi^2 g d^5},$$
(2.3)

This means that the friction factor f is inversely proportional to the Reynolds number Re, according to the well-known expression f = 64/Re.

2. Unstable (partially turbulent) Flow Regime (2000 < Re < 4500). In this range the discharge is unstable and it is difficult to calculate the friction coefficient f, which may change appreciably as a result of minor changes in wall roughness. The flow regime in spiral and labyrinth long-path emitters is of this type, and each type of emitter has a specific pressure-discharge relationship.

3. Turbulent Flow Regime (Re > 4500). For smooth tubes, f is independent of the roughness and decreases with Re. For a rough flow path, if Re > 50000 the friction coefficient has a constant value, depending only on the relative roughness of the flow path, and is independent of Re. This is the case for orifice drippers, for which the discharge of the emitters is proportional to the square root of the pressure (and is therefore less pressure dependent) and may be described by

$$Q = Kd^2 \sqrt{2gH},$$
(2.4)

where K is a coefficient dependent on the type of orifice.

2.1.6
Emitter Pressure – Discharge Relationship

A general empirical formula, valid over a narrow range of operating pressures and characterizing the discharge-pressure relationships of various emitter types, is

$$Q = kH^x,$$
(2.5)

where k and x are constants characteristic of the emitter and of the flow regime in the emitter, respectively. For microtubes with laminar flow, the emitter exponent x = 1, for spiral long-path emitters x = 0.7, for orifice emitters with fully turbulent flow x = 0.5, for pressure-compensated emitters x = 0.0 – 0.1 and for vortex emitters x = 0.4. Values of the emitter exponent x and emitter coefficient k or the pressure-discharge relationships are usually supplied by the manufacturer. Table 2.1 gives some characteristics of several emitters as tested according to the International Standard Organization (ISO). These tests were carried out at CEMAGREF (Centre Nationale du Machinism Agricole, du Genie Rural, des Eaux et des Forets; Bargel et al. 1996; Decroix and Malaval 1985). These data show that in Europe, USA and Israel, accurate emitters with distinct pressure-discharge relations are manufactured, enabling precise irrigation.

Table 2.1. Results of performance tests of emitters at CEMAGREF (Decroix and Malaval 1985; Bargel et al. 1996)

Manu-Facturer	Name	Type	Discharge (Q, l/h)	CV (%)	Pressure (bar)	Coefficient (k)	Exponent (x)
Antelco	Agridrip	On-line	2.35	2.12	1–3	2.29	0.074
Eurodrip	A1	Built-in	3.64	3.65	1–4	2.2	0.157
France	Fani	On-line	2.14	3.08	1.2–4	2.025	−0.041
Irridelco	Flapper	On-line	3.88	11.1	1–3	3.9	0.0051
Hardie	N.G.E.	On-line	1.95	2.14	0.5–3.5	1.83	0.0416
Naan	Tif PC	Built-in	1.99	3.28	1–3.5	1.93	0.0094
Netafim	PCJ	On-line	1.95	2.63	1–4	1.978	−0.007
Lego	Labyrinth	In-line	2.00	6.60	–	–	0.560
Tirosh	Vortex	On-line	8.00	2.1	–	–	0.40
	Longpath	In-line	4.00	2.60	–	–	0.67

2.1.7
Temperature Dependence

In laminar flow, the discharge is inversely proportional to water viscosity [see Eq.(2.2)]; therefore, the discharge rate of emitters with laminar flow have a theoretical temperature dependence of about 2.8%/°C. There are only minor changes in flow due to thermal expansion or contraction of the water passages. The relationship between emitter exponent and temperature effects on discharge is shown in Fig. 2.7 (Decroix and Malaval 1985). Parchomchuk (1976) investigated the temperature effect on different emitter types. He found an increase in discharge of 1.4%/°C for microtubes (x = 1.0) up to a temperature level (depending on pressure and tube diameter) above which the flow became turbulent. Further increase in temperature did not affect the flow. With spiral long-path emitters (x = 0.7) he found a 1.2%/°C increase, up to 29 °C, above which the effect became gradually less (0.7%/°C). The discharge for an orifice-type emitter (x > 0.5) is theoretically temperature-independent, but for various types, Parchomchuk (1976) found a 1–4% increase in flow in the temperature range 7–38 °C. For example, with vortex-type emitters (x < 0.5), he found an 8% decrease in discharge rate with an increase in temperature in the 8–38 °C range. The decrease was probably caused by increased vortex action as viscosity decreased. Similar results were reported by Zur and Tal (1981). Temperature effects may be significant along a lateral. On a bright sunny day, Parchomchuk (1976) found a 16 °C difference between the beginning and end of 20–90 m long laterals with emitters. Such a difference can increase the discharge rate by 11% for spiral long-path emitters, and by 22% for microtubes. This increase in temperature and flow may partially compensate for the pressure loss resulting in flow reduction along the lateral.

2.2
Laterals

Laterals are the tubes on which the emitters are mounted or within which they are inserted or integrated. They are usually made of polyethylene (although PVC

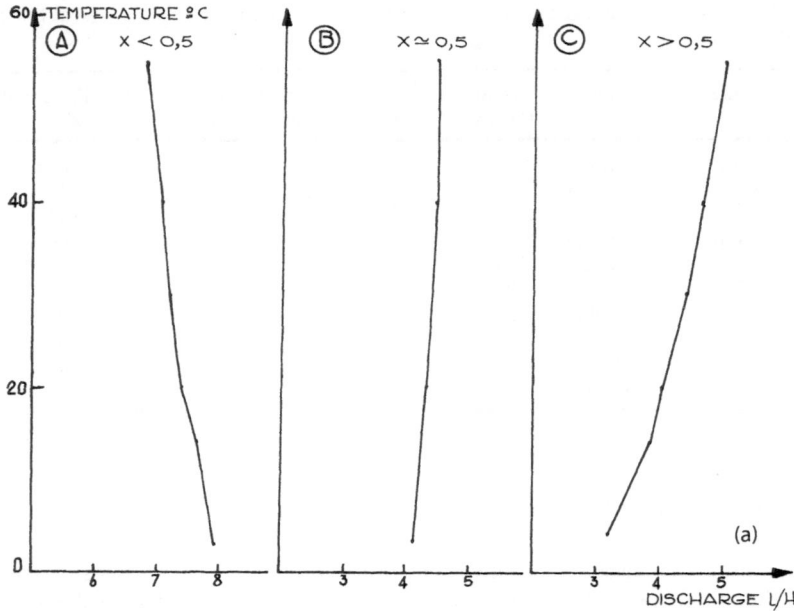

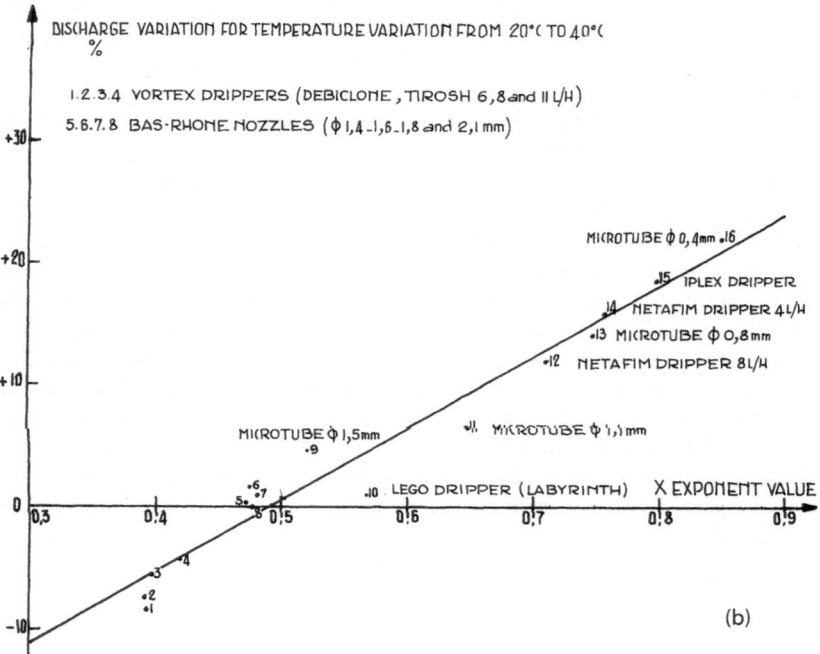

Fig. 2.7. Relative variation of discharge for temperature variation from 20 °C to 40 °C as a function of emitter exponent x (Decroix and Malaval, 1985)

pipes may also be used) with the following features: flexibility, noncorrosivity, resistivity to solar radiation and to the effect of temperature fluctuations, ease of manipulation, and, generally, black in color. (Recently, green or brown-colored laterals have become available, which are more esthetic in landscape irrigation.) Laterals usually have inner diameters in the range of 12–32mm, and wall thickness made to withstand pressure up to 4–6 bar, depending on need. The relationship between the diameter of the tube, the emitter flow rate, the lateral length and the head loss along the lateral is given by empirical formulas. The one most commonly used is that of Hazen-Williams, which may take the form (Bresler 1977):

$$H_L = 2 \cdot 78 \cdot 10^{-6} FLD^{-4.87} (N\overline{Q}/C)^{1.85}, \tag{2.6}$$

where H_L is the head loss in laterals, in m; L is the lateral length, in m; D is the inside diameter of the lateral, in m; N is the number of emitters on the lateral; $\overline{Q}$ is the average emitter discharge along the lateral in m^3/h; C is the Hazen-Williams roughness coefficient (dimensionless), and, according to Howell and Hiller (1974), C = 130 for polyethylene laterals with emitters (D < 16mm), and F is the reduction coefficient to divide flow between emitters along a lateral (dimensionless). For more than 20 emitters per lateral, F = 0.37 (Christiansen 1942).

Various slide rules and nomograms based on Eq. (2.6) are available to facilitate calculations. For example, see Fig. 4.2 in which a log/log plot of head loss versus flow rate is given for different lateral diameters. Howell and Barinas (1980) suggested revising Eq. (2.6) to account for the energy losses across on-line emitter connections. Details on the lateral design are given below (Section 4.3).

2.3
Mains

The main and sub-main lines are usually placed underground and supply water to the laterals. They are normally made of rigid plastic (polyethylene or PVC) in order to minimize corrosion and clogging. The same hydraulic principles as for the laterals [Eq. (2.6)] apply in the design of the main lines. Each lateral connected to the main line can be looked upon as an emitter with a flow rate equal to the sum of the flow rates of all the emitters on the lateral. Details for the design of a complete drip system are given below (Chap. 4).

2.4
Filters

The filter is an essential component of the drip system, its aim being to minimize or prevent emitter clogging. The type of filtration needed depends on water quality and on emitter type (Gilbert et al. 1979, 1981; Ravina et al. 1992). Each type of filter is effective for a particular particle size and type of suspended material, for a specific flow rate, and has a characteristic capacity for sediment collection. In the following we review the principles of operation for the primary types of filters in use.

2.4.1
Centrifugal Sand Separators (Vortex Filters or Hydrocyclone Filters)

The vortex or hydrocyclone filters are effective in filtering sand, fine gravel and other high-density materials from well or river water. Water is introduced tangentially at the top of a cone and creates a circular motion resulting in a centrifugal force, which throws the heavy suspended particles against the walls. The separated particles are collected in the narrow collecting vessel at the bottom (see Fig. 2.8). The collected filtrate (sand) may be emptied manually or by using a special flushing valve. The clean water rises up in a spiral motion to the outlet. The diameters of the top and bottom of the cone must be designed in proportion to the water flow rate. A suitable sand separator can be designed for any water flow rate and sediment load, ranging between 3 and 300 m³/h. With heavy sediment loads, several separators can be installed in a series. The head loss between inlet and outlet is necessary to activate the centrifugal force and ranges 4–15 m, depending on the flow rate. Hydrocyclone filters are particularly effective for primary filtration of surface waters from fast-flowing rivers and canals.

2.4.2
Gravel or Media Filters

Gravel filters are effective in removing light suspended material, such as algae and other organic material, fine sand and silt particles. This type of filtration is essential for primary filtration of irrigation water from open water reservoirs, canals or rivers in which algae may develop. They are also needed for secondary-treated sewage effluent containing suspended organic material. Gravel filters consist of

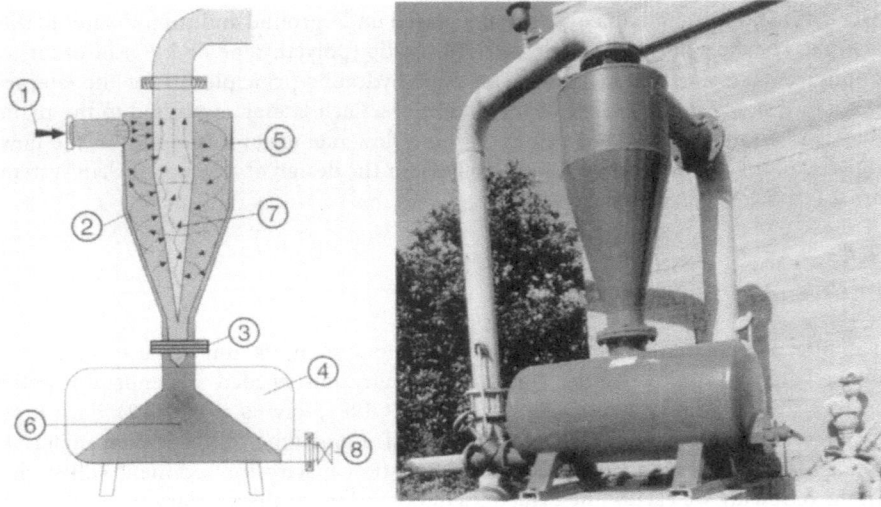

Fig. 2.8. Hydrocyclone filter (By courtesy of NETAFIM Irrigation Equipment and Drip Systems)

fine gravel or coarse quartz sand, free of calcium carbonate (usually 1.5–4 mm in diameter) placed in a cylindrical tank with a diameter ranging between 20 and 200 cm, depending on the capacity of the system. Water is introduced at the top, while a layer of coarser gravel is put near the outlet at the bottom (see Fig. 2.9). The filter is cleaned by reversing the direction of flow and opening the water drainage valve. Pressure gauges are placed at the inlet and outlet of the gravel filter to indicate the condition of the filter. If the filter is clean, the head loss is about 2 m. If the head loss exceeds a certain value, the filter needs cleaning. Automatic self-cleaning filters, based on a preset allowable pressure drop or a set time interval are also available. Media filters enable relatively low flow rates (5–50 m^3/h). Therefore, in many cases a battery of filters connected in parallel is installed at the head of a system.

2.4.3
Screen Filters (Fig. 2.10)

Screen filters are always installed for final filtration as an additional safeguard against clogging. The screens are usually cylindrical and are made of non-corrosive metal or plastic material. Screen filters may be classified as follows:

1. By the diameter of the pipe to which they are fitted (3/4″–4″)
2. By the recommended range of flow rate, which may vary from 3 to 50 m^3/h depending on the diameter (for each screen filter the manufacturer supplies a curve relating flow rate to head loss)
3. By the size of the openings in the screen (in mm, micron or in mesh, i.e., the number of holes per inch). As an approximation, 20, 40, 60, 80, 100, 120, 150 and 200 mesh correspond to screen hole diameters of 0.8, 0.4, 0.25, 0.18, 0.15, 0.13, 0.10 and 0.08 mm, respectively. The most common mesh-size selected for drip irrigation is 100–200 mesh (0.08–0.15 mm in diameter), which is roughly equivalent to 10% of the emitter orifice. Long-term trials with secondary-cleaned wastewater have shown that 80 mesh (0.18 mm) is sufficient for drip irrigation filtration (Ravina et al. 1995)
4. By the total surface area of the filter (in cm^2) and the active or net filter area, which is usually about one third of the total surface area. The ratio between the net filter area and the cross-sectional area of the inlet pipe should be at least 1 : 8
5. By the cleaning method. Manual cleaning by dismantling the screen basket and washing it, back-flushing or draining without dismantling, or automatic back-flushing whenever the head loss across the filter reaches a given magnitude. The head loss across the filter must be measured periodically. If the head loss is higher than the permissible value, according to the manufacturer's specification for the operational flow rate, the filter needs to be cleaned. During cleaning, care should be taken that no foreign material enters the irrigation line.

2.4.4
Disk Filters (Fig. 2.11)

Disk filters contain stacks of grooved, ring-shaped disks that capture debris and are very effective in the filtration of organic material and algae. During the filtration

FILTRATION

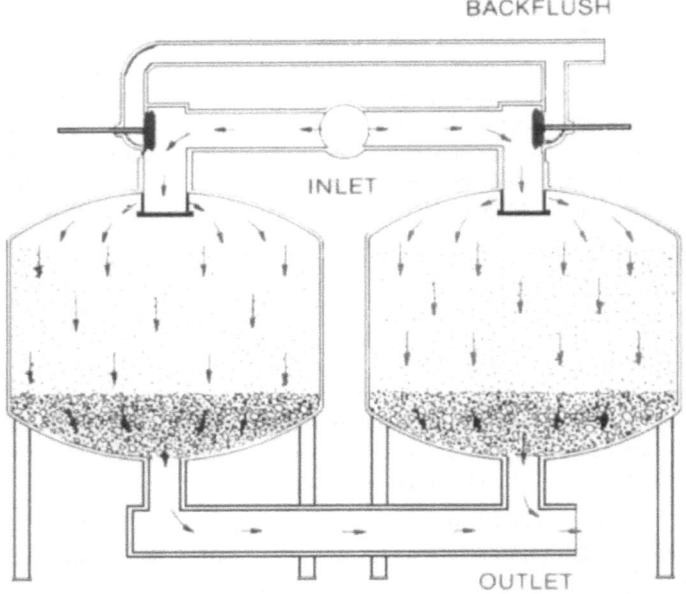

BACKWASHING

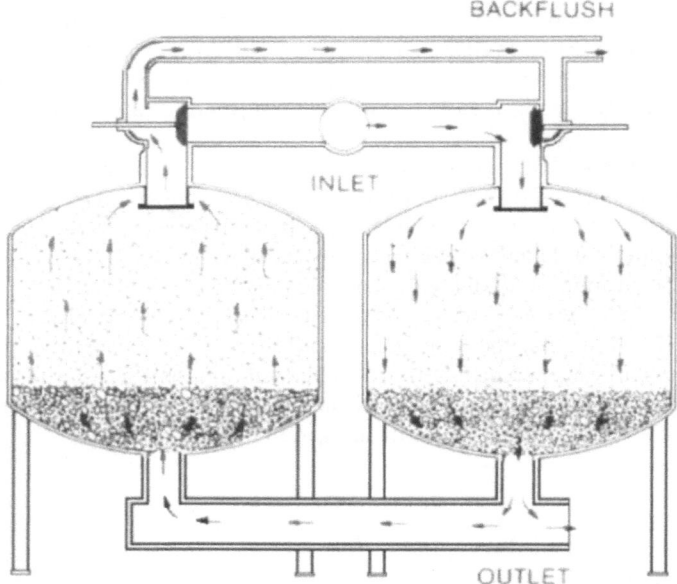

Fig. 2.9. Media filters (By courtesy of YARDNEY Filtration Systems)

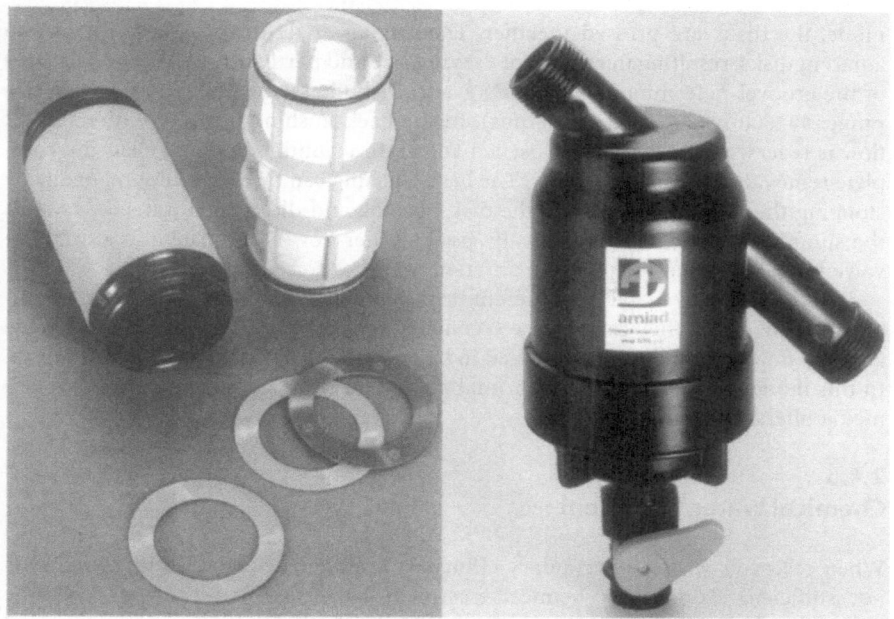

Fig. 2.10. Screen filters (By courtesy of AMIAD Filtration Systems)

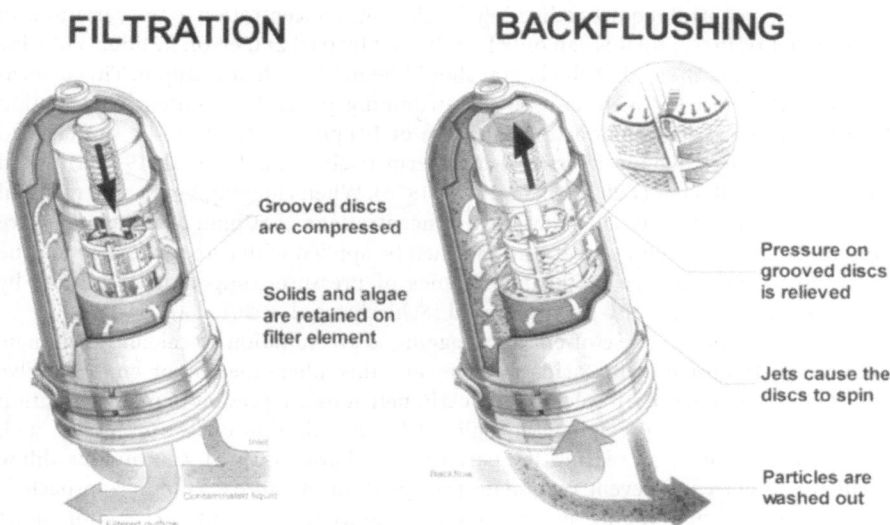

Fig. 2.11. Disk filters (By courtesy of ARKAL Filtration Systems)

mode, the disks are pressed together. There is an angle in the alignment of two adjacent disks, resulting in cavities of varying size and partly turbulent flow. The sizes of the grooves determine the filtration grade. Disk filters are available in a wide size range: 40–600 mesh (400–25 microns). In the back-flushing mode, the direction of flow is reversed. The disks are loosened and start a spinning motion, ensuring complete removal of retained particles. The back-flushing can be carried out manually by stopping the water flow, opening the disk housing and directing a water jet to start the spinning motion, or automatically, as the water flow is reversed and a drainage valve is opened when the pressure difference reaches a certain preset value.

Disk filters are available for different flow rates from 4 to 30 m^3/h (3/4 to 2″ diameter). In this range they can replace secondary screen filters. They are also available at larger flow rates and can be installed in batteries in parallel for more than 500 m^3/h. In this mode they are suitable for primary filtration from reservoirs and can replace media filters.

2.4.5
Chemical Water Treatment

When reservoir water or secondary effluent is used for drip irrigation, filtration is not sufficient. Additional chemical treatment is needed to decompose organic material and to prevent growth of bacterial slimes, algae or protozoa in the irrigation system (Tajirshy et al. 1994; Ravina et al. 1995; Sagi et al. 1995). Some microorganisms and organic material may pass the filters and develop in the laterals and in the narrow water passages of the emitter (Adin and Sacks 1991; Gilbert et al. 1981). This can be prevented by **chlorination**. Chlorine decomposes organic matter, prevents the growth of bacteria, algae and protozoa and prevents the sedimentation of organic suspensions. Chlorine can be applied as chlorine gas, sodium hypochlorite or as calcium hypochlorite. It should be applied at the head of the system, before the filter. Effective chlorination control requires adjusting the chlorine concentration to the variations in water quality. Presently this can only be achieved by trial and error. Residual chlorine concentration at the end of the lateral should be not less than 1–2 ppm. This is measured with a test kit such as used for swimming pools. Intermittent chlorination during 1 h per irrigation at a concentration of 10 ppm was found to be sufficient to prevent organic clogging in several long-term trials (Tajirshy et al. 1994; Hill and Tajirshy, 1995; Ravina et al. 1995; Sagi et al. 1995). When clogging by organic material has occurred, aggressive chlorination at concentrations of 50 ppm or more may solve the problem. These high concentrations must be applied with care, since they may be toxic to plants. Damage to the membranes of pressure-compensated emitters by excessive chloride has also been reported (Schischa et al. 1996).

Another frequent cause of emitter clogging is precipitation of calcium carbonate sediment. **Treatment with acid** may prevent this phenomenon or may dissolve existing precipitates. Both calcium and carbonate ions are present in many irrigation waters. They tend to precipitate at high pH by the addition of fertilizer salts, such as phosphate, or at increasing temperatures (Nakayama 1986). Continuous dilute acid treatment can prevent carbonate precipitation. A more common approach is to dissolve precipitates by using high acid concentrations (at pH 2) for short periods. Hydrochloric, sulfuric and phosphoric acid may all be used. Phosphoric acid is preferable, since it is the least corrosive and has plant nutritional value. Acid should be added after the filters, in order to prevent corrosion damage to the equipment. The ends of the laterals should be opened for flushing after approximately 1 h of acid treatment.

Herbicides may be added to subsurface drip irrigation systems, depending on environmental regulations and special certification, in order to prevent root intrusion into the emitters. Treflane is usually employed since it is strongly adsorbed into the soil and stays in the vicinity of the emitter (Gerstl et al. 1981). Treflane-impregnated emitters which release the herbicide slowly into the soil are also available (e.g., Geoflow, Rootguard). Continuous application of acid as used in the prevention of precipitates may also serve as a herbicide (Howell et al. 1997). **Bactericides**, apart from chlorine, are also used to prevent the development of bacterial slime in the laterals and emitters. This is especially important when treated effluent is used in irrigation. Slow-release bactericides incorporated in the laterals are now available (e.g., Geoflow, Wasteflow).

2.5
Fertilizing Systems

The fertilizing systems used to add chemicals (nutrients, herbicides or pesticides) to the irrigation water are considered an integral part of the drip system. The process of adding fertilizers to the drip system is called "fertigation". The importance of fertilizing in conjunction with the drip method was described above. However, fertigation is not free of hazards. The chemicals added to the water may be toxic to humans or animals. Consequently, safeguards must be taken to prevent back-flow of irrigation water into a main water supply system that might be used for drinking water. Check valves are used for this purpose. Non-corrosive material should be used for the fertilizer containers and injection equipment. Special fertilizers are used, as described above, to reduce the potential for clogging by precipitation of impurities or by salts formed in chemical reactions between fertilizers and salts present in the irrigation water. Several methods of fertigation are available.

2.5.1
Venturi Tube Principle

A constriction in the main water flow pipe increases the water flow velocity thereby causing a pressure differential ("vacuum") which is sufficient to suck fertilizer solution from an open reservoir into the water stream. (Fig. 2.12A). The rate of flow can be regulated by means of valves. This is a simple and relatively inexpensive method of fertilizer application, but it has some disadvantages: the pressure loss across a venturi valve is high, about one third of the operating pressure. Moreover, precise regulation of flow is difficult because the rate of injection is very sensitive to the pressure and rate of flow in the system. By installing a Venturi valve on a by-pass of the total irrigation flow, head loss is decreased considerably and the fertilizing system can be disconnected and moved easily.

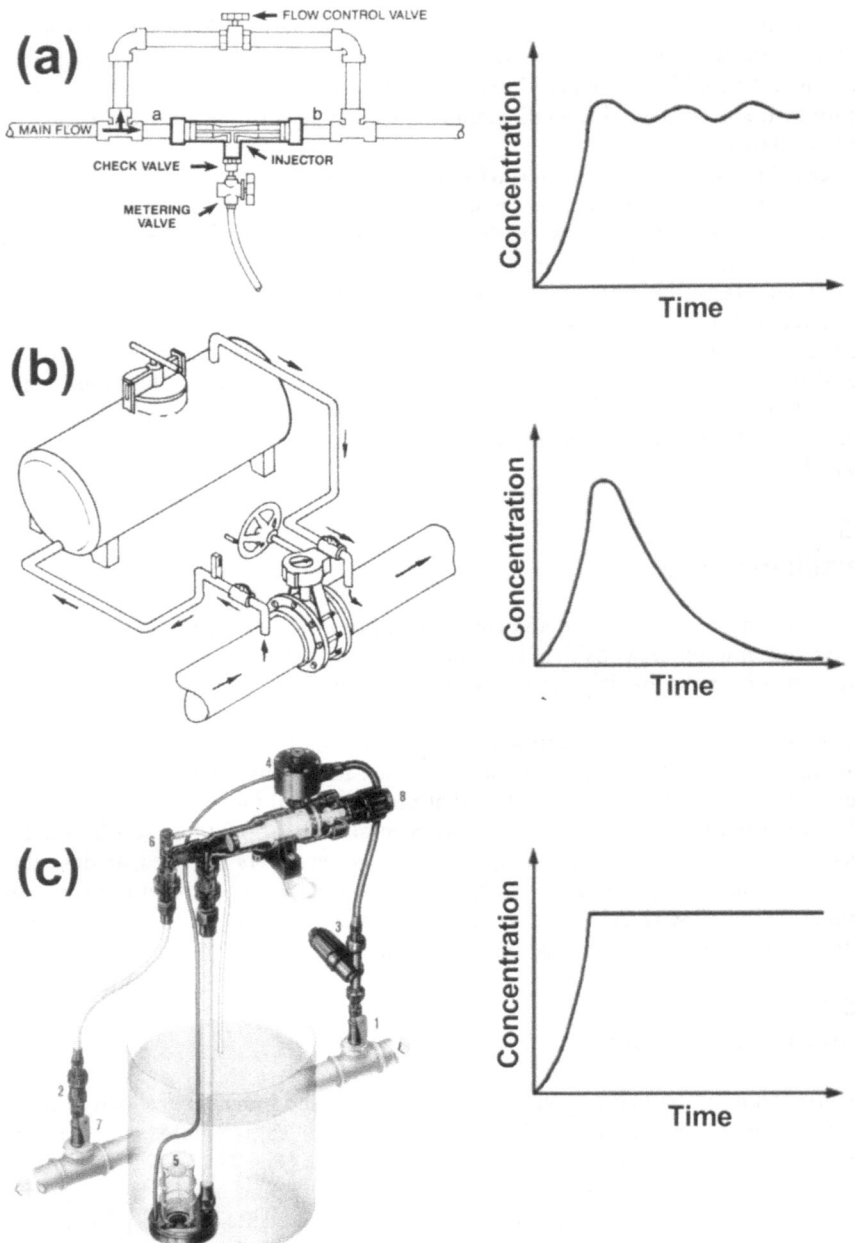

Fig. 2.12. Fertilizer injection systems. **a**: Venturi (MAZZEI Injector Cooperation). **b**: Fertilizer tank (T SYSTEMS International). **c**: Hydraulic injection pump (AMIAD Filtration Systems)

2.5.2
Fertilizer Tank (By-Pass System)

This method employs a tank into which the dry or liquid fertilizer is placed. The tank is connected to the main irrigation line by means of a by-pass so that some of the irrigation water flows through the tank and dilutes the fertilizer solution. This by-pass flow is brought about by a pressure gradient between the entrance and exit of the tank, created by a permanent constriction in the line or by a control valve (Fig. 2.12B). The concentration of fertilizer in the tank decreases gradually, until it reaches the level of the irrigation water, according to the following equation (Bar Yosef 1977):

$$C(t) = C_o \cdot \frac{J}{J_T} \cdot e^{\left(-\frac{J}{V}t\right)}, \tag{2.7}$$

where $C(t)$ is the changing solution concentration with time, J is the water flow in the main line to the field and J_T is the fertilizer flow rate, V is the volume of the tank and C_o is the initial concentration in the tank at t = 0. Experience has shown that with liquid fertilizers it takes approximately four tank-volume displacements to empty the tank of fertilizer. If solid fertilizer is used, at least ten volume displacements are needed to dissolve all the material. The rate of flow through the by-pass is determined by the pressure head difference between entrance and exit, which is usually in the order of 1–5 m water. The choice of tank size (available sizes, 50–1000 l) is related to the area being irrigated. The pressure difference needed in order to gradually empty the tank during one irrigation has to be determined empirically.

The advantages of this system include: (1) simplicity in construction and operation; (2) low cost; (3) no external power supply needed; and (4) relative insensitivity to changes in pressure or flow rate. However, the tank must be strong enough to withstand the pressure of the irrigation line. The disadvantages of the system include the following: (1) The varying concentration of nutrients causes the bulk of the fertilizer to be applied at the beginning of the irrigation cycle; (2) The tank must be refilled with solution for each irrigation cycle; thus the system is not suitable for automatic or serial irrigation.

Nevertheless, because of its simplicity and low cost, this is the most common fertilizer system in use in semi-automatic drip installations for small areas (e.g., vegetables, flowers).

A variation of this system uses a rubber diaphragm in the tank to separate the fertilizer solution from the irrigation water. The rate of flow of the solution is determined by the pressure difference between the inlet and outlet points and is monitored by a flow meter. This technique ensures a relatively constant fertilizer supply rate to the irrigation water, provided the flow rate of the fertilizer solution is adjusted to the flow rate of the irrigation water.

2.5.3
Injection System

With this method a pump is used to inject fertilizer solution into the irrigation line. The solution is normally pumped from an unpressurized reservoir, and the choice of pump type used is dependent on the power source. The pump may be driven by an internal combustion engine, an electric motor, a tractor power takeoff, or hydraulic pressure ("water engine"). The electric pump can be automatically controlled and is thus the most convenient to use. However, its use is limited by the availability of electrical power, and is therefore more suited to glasshouse than to field operation. With these injection devices, fertilizers may be supplied to the irrigation water at a more or less constant rate. The pumping rate and the concentration of the stock solution can be adjusted to attain the desired level of fertilizer application. However, the water flow and the fertilizer flow are independently controlled. Changes in water flow rate, power failure or mechanical failure may cause serious deviations from planned concentrations. Another disadvantage of this system is the need for an external power source and the relatively high cost of the system. The use of a hydraulic engine, operated by the line pressure, avoids these difficulties (see Fig. 2.12C). This device requires a minimum pressure of usually 10–15 m water to operate. At each stroke of the pump a certain predetermined volume of fertilizer solution is injected into the irrigation system. The number of strokes is determined by the line pressure. The hydraulic engine may be activated by a pulse-generating water meter so that the fertilizer injected is exactly proportional to the water flow and a constant concentration is maintained. Automatic computerized control systems are also available to provide exact injection of fertilizer into the water flow.

2.6
Automation

Drip irrigation enables a large degree of control of water application in a well-designed system. Water application is precise with a high degree of uniformity between emitters. It is possible, therefore, to apply water according to the exact demands of the plant as determined by changes in weather, crop growth stage and desired plant water stress, and soil water salinity. The amounts of fertilizer applied can also be adjusted to demand, according to plant growth stage. Automation is employed to control the system's main valves in order to determine irrigation amounts. Two basic control systems are available – one based on timing and the other on water flow measurements. Timing devices are based on the assumption that the rate of flow in the system is constant; for a given time, a constant volume of water passes the valve, which is controlled electrically or hydraulically. Such conditions do not always exist since changes in water pressure occurring in the system due to other users may cause changes in water flow. Therefore, the second system based on direct water flow measurements is considered more accurate.

Semi-automatic systems are based on manually setting a control valve to a predetermined time or to a predetermined amount of water. These devices are of limited value since each valve has to be preset before each irrigation for each section of the field. They are best used in small plots of vegetables or flowers and in landscape irrigation.

Fully automatic systems enable the control of several valves, coupled with fertilizer injection systems, flushing of filters and input from sensors monitoring soil water status, plant water status and evaporative demand as will be detailed in Chapter 5.

The introduction of more sophisticated controllers and later on general-purpose computers for irrigation control and management requires feedback for control, performance assessment, and decision making. A host of transducers for valve control, pressure, flow and liquid level monitoring as well as other types of measurements are now being incorporated into modern drip irrigation systems. Most of these components are designed to provide either analog or digital output of the state of the component. Some of more common elements are described below:

Flow Meters. Most common flow meters used in irrigation are anemometer and propeller velocity types (others types such as electromagnetic velocity, Doppler and optical strobe velocity are also available). These flow meters use anemometer cup wheels or propellers to sense water velocity. Most modern units are capable of providing a calibrated electrical output (as digital counts per unit volume). Total amounts of flow and flow rates are provided by analog gauges, or may be calculated from the digital output with the aid of a controller or a computer. Some water meters are combined with a dosing mechanism and a valve to serve as a semi-automatic control device. A unit equipped with an impeller for flow measurements and a control head is depicted in Fig. 2.13.

Flow meters are extremely important elements for verifying irrigation amounts applied, for monitoring proper performance and to provide early indication of emitter plugging.

Automatic (Hydraulic/Electric) Valves. Remotely controlled electric and hydraulic valves provide the means for executing irrigation and fertigation scheduling, as well as enabling automatic filter maintenance. Valves are activated by electric or hydraulic

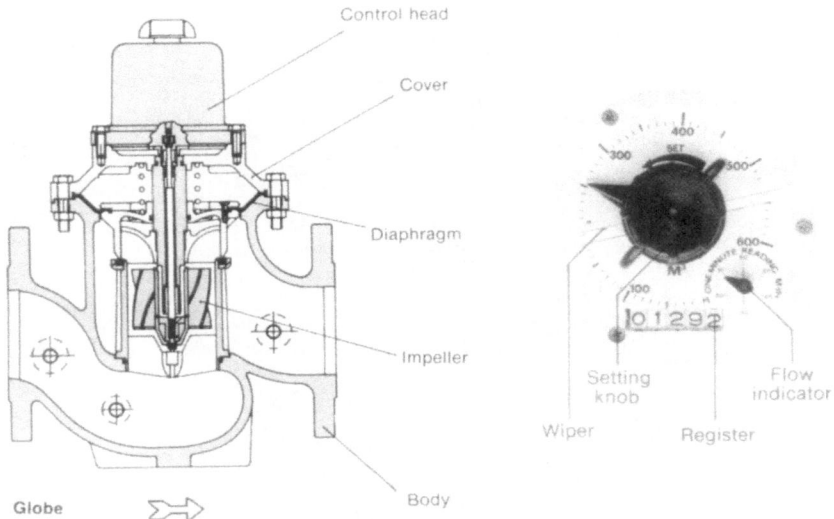

Fig. 2.13. Semi-automatic metering valve (By courtesy of BERMAD Control Valves Inc)

signals sent by a controller through a solenoid on the valve, or by directly pressurizing or evacuating a valve's hydraulic chamber. An example of an electric valve is depicted in Fig. 2.14.

Advanced methods for addressing and controlling valves (other than using dedicated wires) often control many valves using a single cable. Each valve is equipped with a circuit board with an individual addressing mechanism (similar to addressing through PC parallel or serial ports). More recently, technology similar to that used in

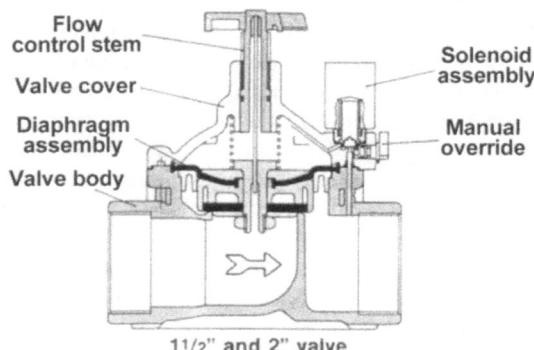

Fig. 2.14. A diagram of an electric valve (By courtesy of BERMAD Control Valves Inc)

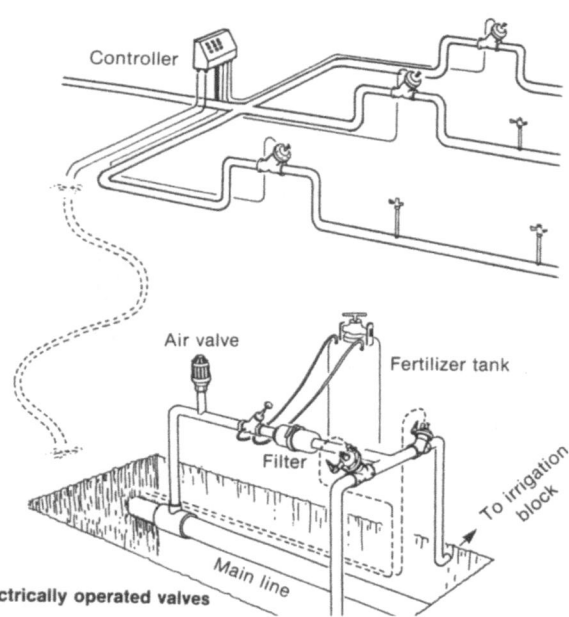

Fig. 2.15. A simple application controller (By courtesy of BERMAD Control Valves Inc)

personal paging systems is being used to remotely address stand-alone valves (known as "radio valves").

Pressure Gauges and Transducers. Many automatic filtration batteries are equipped with mechanical or electric pressure transducers that gauge the pressure difference across the filters and often provide information for initiation of an automatic backflush sequence. Such a setup often requires electric or hydraulically controlled valves, a pressure sensing unit and a controller. An automatic setup offers a distinct advantage over a timer (where backflushing is performed at preset time intervals) when water quality and flow volumes vary during an irrigation cycle.

Controller: A controller receives and integrates information from various sources (flow meters, pressure transducers, water level meters, soil water sensors) and is capable of issuing commands to various valves to initiate or terminate irrigation or fertilizer injection. This can be based on a preset program or on the occurrence of abnormal flow rates. Safeguards can be installed to shut down the system and to alarm the irrigator if the pressure is too low or if a leak has developed. The advances in computers and processors are reflected in the broad tasks that an average controller is capable of performing (see Fig. 2.15)

Soil Water and Salt Regime

3.1
Modeling Soil Water Regimes

3.1.1
Water Flow Equations

Most traditional irrigation methods are designed to provide uniform water distribution at the soil surface resulting in wetting patterns similar to those under natural rainfall. Under these circumstances, the predominant force governing movement of water in the soil during irrigation is gravity. During the subsequent evaporation and transpiration period, capillary forces become dominant and the net movement of water is in a vertically upward direction. Hence, the infiltration of water into the soil and subsequent redistribution and evaporation can be analyzed, mainly as a one-dimensional flow process which obeys the Richards' equation governing water flow:

$$\frac{\partial \theta}{\partial t} = \frac{\partial}{\partial z}\left[K(\theta)\frac{\partial H}{\partial z}\right]. \tag{3.1}$$

This water flow equation states that temporal changes in water content at any given point in the soil are equal to the concurrent changes in specific flux q in the vertical direction z. This flux is defined by Darcy's law as: $q = -K(\theta)\ \partial H/\partial z$, where $K(\theta)$ is the unsaturated hydraulic conductivity, which is a function of the water content, and $\partial H/\partial z$ is the hydraulic head gradient. The hydraulic head H is the sum of the gravitational head z and the matric (pressure) head h (i.e., $H = h + z$).

Water flow under drip irrigation is more complex, since water is applied from emitters (which may be considered as small circular or point sources), and from each emitter the water spreads in all directions. The resulting water infiltration process is three-dimensional with respect to the three space coordinates. If emitters are sufficiently far apart, we can regard each emitter as an independent unit, wetting its own soil volume, without interactions with neighboring emitters. A useful representation of flow processes from emitters is based on an assumed cylindrical geometry in which water flows from an emitter into a cylindrical soil element of radius R (2R being the spacing between two adjacent emitters), and depth Z, chosen to be larger than the depth of wetting and depth of the root zone. Assuming uniform soil conditions in the cylindrical soil element, we can regard the infiltration process as an axisymmetric two-dimensional flow problem, with radial symmetry. The water flow equation therefore takes the form:

$$\frac{\partial \theta}{\partial t} = \frac{\partial}{\partial r}\left[K(\theta)\frac{\partial h}{\partial r}\right] + \frac{K(\theta)}{r}\frac{\partial h}{\partial r} + \frac{\partial}{\partial z}\left[K(\theta)\frac{\partial H}{\partial z}\right], \tag{3.2}$$

where r and z are the radial and vertical dimensions, and h is the soil-water matric (pressure) head. Another but common flow configuration results when emitters are closely-spaced on a line such that their wetting patterns overlap, or for a long perforated (or porous) tube. If we let the line be aligned with the y axis (with x, the horizontal coordinate normal to y, and z the vertical coordinate), the flow becomes independent of y (plane flow). The governing water flow equation for plane flow (also known as flow from a line source) is:

$$\frac{\partial \theta}{\partial t} = \frac{\partial}{\partial x}\left[K(\theta)\frac{\partial H}{\partial x}\right] + \frac{\partial}{\partial z}\left[K(\theta)\frac{\partial H}{\partial z}\right]. \tag{3.3}$$

Solutions for Eqs. (3.2) and (3.3), subject to realistic initial and boundary conditions, provide important information for designing drip systems which take into account soil and emitter properties.

The representation of water flow processes by equation (3.2) or (3.3) lacks information on the role of water uptake by plant roots on water distribution. This information is essential for proper design and management of drip systems under cropped conditions, especially in situations where irrigation scheduling is based on monitoring soil water status using soil water sensors or direct soil sampling. A more general mathematical representation of multidimensional water flow, considering water uptake by plant roots, is given (in vector notation) by:

$$\frac{\partial \theta}{\partial t} = \nabla \cdot [K(\theta)\nabla H] - S, \tag{3.4}$$

where ∇ is the gradient operator, and S represents plant water uptake as a volumetric sink term (volume of water taken up by plant roots per volume of soil per unit time). The challenge to practical application of Eq. (3.4) is the determination of the functional form of S which describes both the spatial distribution and intensity of uptake for a particular crop (a discussion of these considerations follows).

The motivation for seeking solutions to water distribution from emitters [i.e., solutions for Eqs. (3.2) to (3.4)] is driven by design and management needs. For example, an important drip system design parameter is the spacing between emitters for a given discharge rate and soil type, which may be calculated for known soil hydraulic properties (K and h as functions of θ). Another example is the development of general guidelines for soil water sensor placement within the crop root zone (for soil water status monitoring) applicable for various crops, soils, and sensor types. In the following sections we will discuss several methods for obtaining such solutions.

3.1.2
Numerical Solutions – General Flow Conditions

The Richards water flow equation (Eqs. (3.1) to (3.4)) is difficult to solve due to the highly nonlinear dependency of K on the unknown water content θ. Thus, except for a few special situations, Richards' equation can be solved only by numerical methods such as finite differences (Brandt et al. 1971; Lafolie et al. 1989) or finite elements

(Taghavi et al. 1984). One of the earliest and most comprehensive solutions to the problem of water flow from an emitter placed on the soil surface was developed by Brandt et al. (1971). They assumed the soil to be a stable, isotropic, homogeneous, porous medium. The initial water content (i.e., initial conditions) was assumed to be uniform and low (relatively dry soil). The emitter is placed on the soil surface at the center of a circle of radius R equal to half the distance between emitters and has a certain known volumetric flow rate or discharge q. Based on observations that in the vicinity of the emitter a radial area of ponded water develops, and that the water content at the soil surface beneath this ponded area is at saturation ($\theta = \theta_s$), it was assumed by Brandt et al. (1971) that water infiltrates into the soil only through this ponded-saturated area.

The details of the numerical procedure and computer scheme for solving water-content distribution and the water fluxes in the soil are beyond the scope of this book. However, comparisons of computed soil-water content data and the data obtained from laboratory and field studies verify the validity of the model (Bresler et al. 1971; Bresler and Russo 1975). Some data from this work are given in Fig. 3.1 and show good agreement between calculated and experimental data. Good agreement was also obtained for field data. In all these comparisons the surface evaporation was taken to be negligible. However, even a high evaporation rate (10 mm/day) would not have an appreciable effect on water-content distribution (Bresler 1975) since the rate of evaporation is normally low compared to the soil saturated hydraulic conductivity (1–10 cm/h). Consequently, only when evaporation is extremely high, or when the hydraulic conductivity of the soil is very low (<0.1 cm/h), does one need to consider the effect of evaporation during the infiltration phase of drip irrigation.

The effect of emitter discharge rate and saturated hydraulic conductivity of the soil on the radius of the saturated zone and on the wetting front advance is illustrated in Figs. 3.2 and 3.3, respectively (Bresler 1978). The ability to accommodate the (constant) emitter discharge diminishes as the underlying soil volume wets, resulting in an increase in the radius $\rho(t)$ of the ponded entry zone with time t. The rate of increase in the radius of the saturated zone decreases with time and reaches a maximum. The maximum radius depends on the soil saturated hydraulic conductivity and is strongly affected by the discharge rate of the emitters. In practice, this means that in heavy-textured soils (lower hydraulic conductivity), and at higher discharge rates, the distance between emitters can be larger. The figures also clearly show that the shape of the wetted zone is affected by emitter discharge rate and by soil hydraulic properties. The above information is useful for design purposes, as emitter discharge and distances between emitters can be adjusted to the soil hydraulic properties.

Levin et al. (1979) and, more recently, Pal et al. (1992) obtained good agreement between measurements and calculated values from Bresler's (1975) soil-water distribution patterns model. Though the numerical models and experiments of Brandt et al. (1971) and Bresler (1975) remain the most quoted, many computational improvements and other refinements were developed over the years (e.g., Lafolie et al. 1989; Taghavi et al. 1984). Simunek et al. (1993) developed a general-purpose and user-friendly two-dimensional numerical model (known as SWMS 2D, or more recently as Hydrus-2D). It is based on a finite-elements scheme and incorporates a graphical user-interface for data entry (soil properties, atmospheric conditions, emitter discharge, boundary conditions, etc.), and provides visualization of simulation results. The present trend of developing general-purpose simulation models (such as Hydrus-2D)

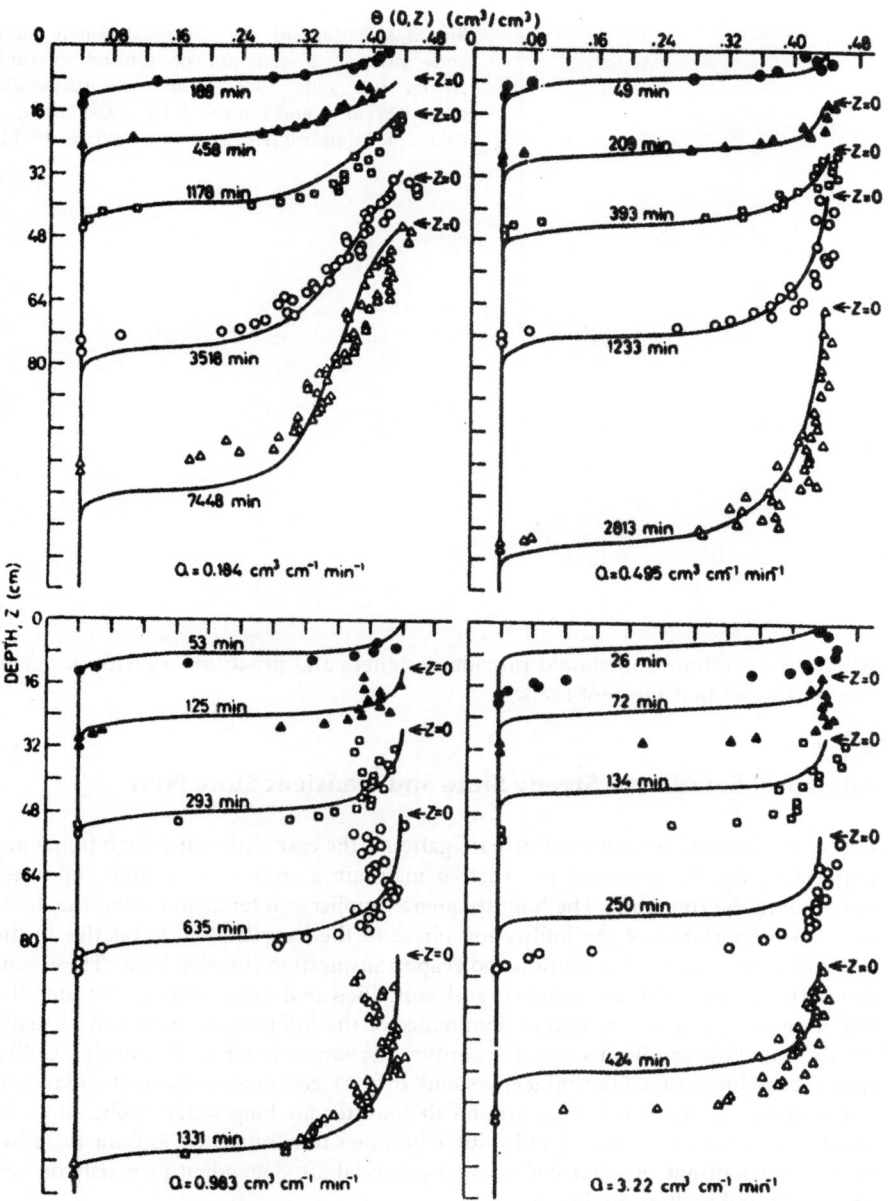

Fig. 3.1. Vertical water content distribution θ(0,z) in the plane of symmetry (x = 0). Computed (*lines*) compared with experimental (*points*) data of Gilat loam soil. The numbers indicate infiltration time (t), Q = discharge per unit length. Note that the (z = 0) plane is shifted (Bresler et al. 1971)

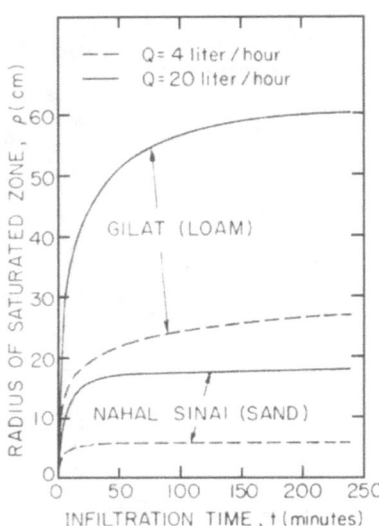

Fig. 3.2. Radius of the saturated water entry zone $\rho(t)$ as a function of infiltration time (t) for two soils (loam with Ks = 0.84 cm/h, $\alpha = 0.025\,\text{cm}^{-1}$; and sand with Ks = 8.4 cm/h, $\alpha = 0.065\,\text{cm}^{-1}$) and two discharge rates (Bresler 1978)

is likely to continue and should provide designers and practitioners with powerful diagnostic and management tools.

3.1.3
Analytical Solutions – Steady State and Transient State Flow

One of the important features of drip irrigation is the ease with which high frequency application can be practiced in order to maintain a continuously high soil-water potential in the root zone. The high frequency application tends to extend the duration and importance of the infiltration phase in the irrigation cycle relative to the other phases such as redistribution and evapotranspiration (Bresler, 1978). These conditions have motivated researchers to seek simplified analytical solutions to water distribution from emitters assuming dominance of the infiltration phase and virtually steady-state flow conditions (i.e., the emitter discharges water continuously). Unlike flow in one-dimensional (sprinkler) systems, drip irrigation water content and potential assume a finite distribution around the emitter for long water applications. In other words, not all the wetted soil volume becomes fully saturated, and the resulting spatial distributions of water content and potential are dependent primarily on soil properties and emitter discharge.

The derivation of analytical solutions to the Richards equation, under steady-state flow conditions, requires several simplifying assumptions which limit their applicability to real field conditions. Nevertheless, several advantages of analytical solutions over numerical solutions include: (1) modest parameter requirements; (2) provision of insight and a direct link between input parameters and resultant soil water conditions; and (3) provision of a general framework which facilitates design formulation and management guidelines.

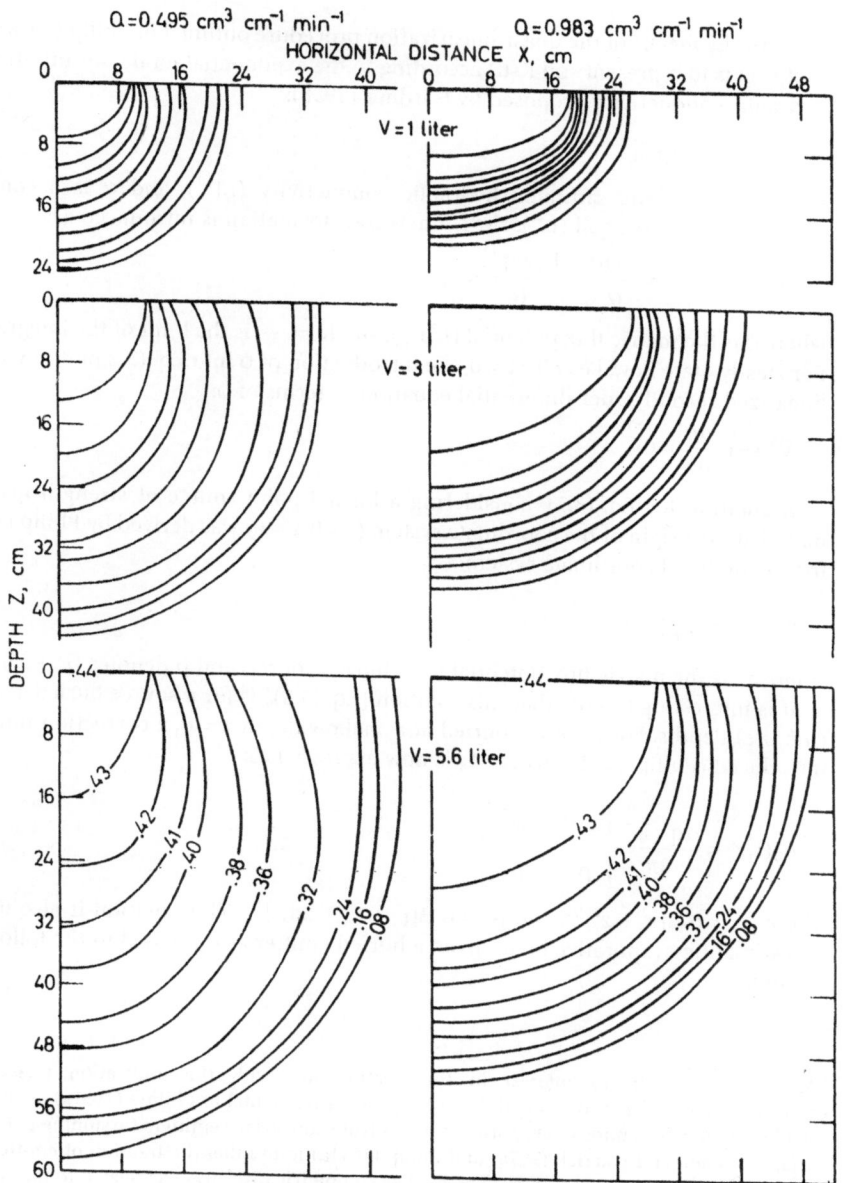

Fig. 3.3. Water content as a function of cumulative infiltration V for two cases of discharge Q. The numbers on the curves indicate water content θ (Brandt et al. 1971)

For illustration purposes we consider the multidimensional infiltration described by Richards' equation without plant uptake, focus on flow from emitters represented as point sources only (surface and subsurface) and ignore other possible source configurations such as line sources [Eq. (3.3)]. Analytical solutions for Eq. (3.2) are obtained by means of the quasi-linearization procedure outlined in Philip (1968). The first step is to represent soil K(h) according to the exponential model for unsaturated hydraulic conductivity proposed by Gardner (1958):

$$K(h) = K_s \exp^{\alpha h}, \tag{3.5}$$

in which K_s is the saturated hydraulic conductivity (LT^{-1}), and α is a constant characteristic of the soil (L^{-1}). Next, a new transformation is introduced as:

$$\phi = \int_{-\infty}^{h} K(h)\,dh = \frac{K(h)}{\alpha} = \frac{K_s \exp^{\alpha h}}{\alpha}, \tag{3.6}$$

where ϕ is the matric flux potential ($L^2 T^{-1}$), and $h_o = -\infty$ is the limit of the integration. For steady-state flow, i.e., $\partial\theta/\partial t = 0$, the introduction of ϕ into Eq. (3.2 or 3.3) yields a linearized second-order differential equation in terms of ϕ:

$$\nabla^2\phi - \alpha\frac{\partial\phi}{\partial z} = 0. \tag{3.7}$$

A solution for Eq. (3.7), considering a buried point source of strength $q[L^3 T^{-1}]$ placed at the origin of the coordinate system ($r = 0, z = 0$) was derived by Philip (1968) for a cylindrical coordinate system as:

$$\phi_B(r,z) = \frac{q}{4\pi\rho}\exp^{\frac{\alpha}{2}(z-p)}, \tag{3.8}$$

where ϕ_B is the matric flux potential for a buried source, and ρ denotes $(r^2 + z^2)^{1/2}$.

It is interesting to note that this solution [Eq. (3.8)] is for a source buried at great (infinite) depths. For a source buried at a shallow depth $z = d$, a correction must be introduced (Philip, 1991a) where ϕ_B is now expressed as:

$$\phi_B(r,z) = \frac{\alpha q}{8\pi}e^{\frac{\alpha z}{2}}\left[\frac{e^{-\rho^*}}{\rho^*} - \frac{e^{\rho_1^*}}{\rho_1^*}\right], \tag{3.9}$$

where $\rho^* = \alpha/2[r^2 + z^2]^{1/2}$, and $\rho_1^* = \alpha/2[r^2 + (z + 2d)^2]^{1/2}$. This solution is also useful for estimating evaporative losses from a buried emitter as discussed in the following footnote[1].

---

[1] Although our primary interest in this section lies with the infiltration phase, the corrected solution for emitter depth [eq. (3.9)] was instrumental in Philip's (1991a) analysis of establishing upper bounds to evaporation losses from subsurface emitters (assuming a dry soil surface). Lomen and Warrick (1978) and Philip (1991a) found that the fraction of emitter discharge being lost to evaporation E (expressed as volumetric discharge), is related to the emitter depth d, and to the soil sorptive parameter α, through the simple equation (Lomen and Warrick 1978; Philip 1991a):

$$\frac{E}{q} = \left(\frac{m}{2+m}\right)\exp^{-d\alpha}. \tag{3.10}$$

Philip (1991) considers the constant $m = \infty$, whereas Lomen and Warrick (1978) calculate m from the maximum evaporative demand from a wet soil E_s (expressed as evaporative flux per unit area) as: $m = 2E_s/K_s$. Lomen and Warrick (1978) also considered evaporative losses under transient flow conditions and from surface sources.

Raats (1972) derived an analytical solution to Eq. (3.7) for a point source on the soil surface:

$$\Phi_s(r,z) = 2\Phi_B + \frac{\alpha q \, e^{\alpha z}}{4\pi} E_i \left[\frac{\alpha}{2} \left(z + (r^2 + z^2)^{1/2}\right) \right], \tag{3.11}$$

where Φ_S and Φ_B are the dimensionless matric flux potentials for surface and buried sources, respectively, defined as $\Phi = 8\pi\phi/\alpha q$ [note that ϕ_B is calculated from Eq. (3.8)],

and E_i is the exponential integral, $E_i(x) = \int_x^{\infty} \dfrac{e^{-x}}{x} dx$ (Abramowitz and Stegun 1964).

These two analytical solutions [Eqs (3.8) and (3.11)] can be used to predict the steady-state distribution of the soil matric potential h and soil water content θ, (when h-θ relationships for the soil are known) for surface or buried emitters. The conversion of matric flux potential ϕ to matric potential h, follows from eq. (3.6) as:

$$h(r,z) = \frac{1}{\alpha} \ln\left[\frac{\alpha\phi(r,z)}{K_s} \right]. \tag{3.12}$$

A comparison of theoretical and experimental distributions of soil matric potential for surface and subsurface emitters is depicted in Fig. 3.4 (Coelho and Or 1997).

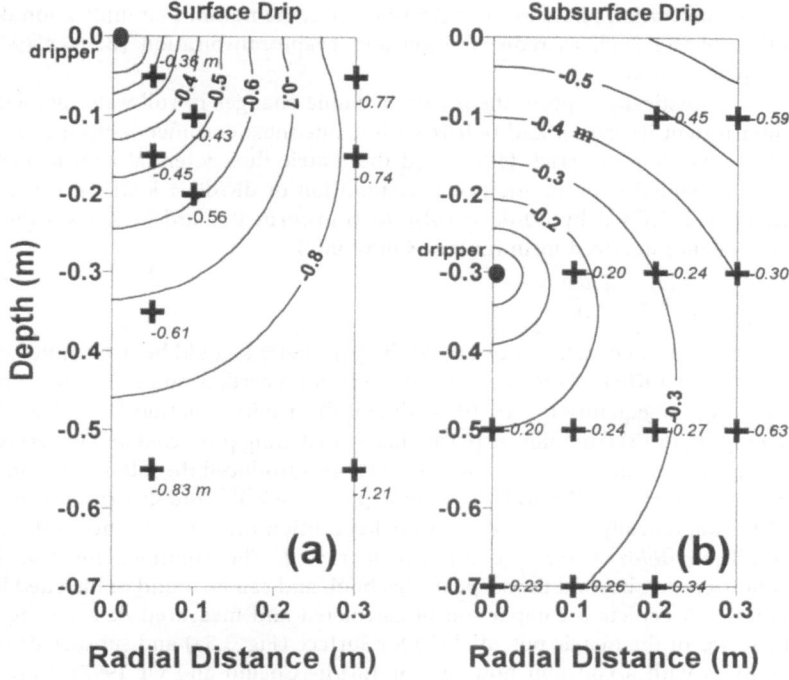

Fig. 3.4. Predicted and obseved matric potentials at steady-state obtained by analytical solution for **a:** surface point source [Eq. (3.11)] for Milville silt loam initially air dry at a dripper flow rate of 0.43 l/h and **b:** buried point source [Eq. (3.8)] with initial water content = 0.18 m³/m³ at a flow rate of 1.6 l/h (Coelho and Or 1997)

In practice, most surface emitters produce a small pond during the infiltration phase of irrigation, especially for high flow rates and soils with low intake properties. The radius of the pond r_s will reach a maximum size given by (Wooding 1968; Bresler 1978):

$$r_s = \sqrt{\frac{4}{\alpha^2\pi^2} + \frac{q}{\pi K_s} - \frac{2}{\alpha\pi}}. \qquad (3.13)$$

Bresler (1978) and others have used steady-state flow solutions based on Wooding's (1968) analysis and Eq. (3.13) for developing design criteria for emitter spacing. Though the basic premise behind relating emitter-soil flow interactions of a drip system is sound, practical applications of the results are lacking. An important question is, how realistic is the assumption of steady-state for real field applications of drip irrigation? One can design systems with very high irrigation frequencies in which soil water content and matric potential about the emitters may reach near steady-state values (as shown by Jury and Earl (1977)) for three irrigations per day. However, under most practical conditions found in irrigated fields, even for a daily irrigation, the temporal variations in h and θ may be large and not resemble their theoretical steady-state values even at positions close to the emitter. The duration of the infiltration phase is often about one-tenth of an irrigation cycle. This was confirmed recently by Coelho and Or (1997) who found that though high frequency application tends to extend the duration and importance of the infiltration phase in the irrigation cycle, the notion that infiltration dominates other phases such as redistribution and evapotranspiration (Bresler 1978) rarely happens in practice.

To realistically capture the highly dynamic changes in soil water associated with intermittent irrigation and redistribution, one must consider solutions for transient flow conditions. Warrick (1974 used the matric flux potential transformation [Eq. (3.6)], coupled with an additional assumption of $dK/d\theta = k$ where k is a constant (it is also defined by $d\theta/d\phi = \alpha/k$), to transform Richards' equation into a time-dependent linearized form (in terms of ϕ only):

$$\frac{\partial\phi}{\partial t} = \frac{k}{\alpha}\nabla^2\phi - k\frac{\partial\phi}{\partial z}. \qquad (3.14)$$

According to Ben-Asher et al. (1978) the parameter k could be determined either: (1) from known $K(\theta)$ relationships for the soil (for a certain range of water content); (2) by using the equality $k = \alpha D(\theta)$ with soil diffusivity function D, evaluated at some average θ; or (3) from flow experiments by matching predicted and observed h and θ using k as a fitting parameter. Warrick (1974) introduced the following dimensionless variables: $R = \alpha r/2$, $Z = \alpha z/2$, $T = \alpha kt/4$, $\rho = (R^2 + Z^2)^{1/2}$, and $\Phi = \alpha q\phi/8\pi$, to solve Eq. (3.14) analytically, subject to the initial condition $\phi(r,z,0) = 0$, and to the boundary condition $\partial\phi/\partial z + \alpha\phi = 0$ for $z = 0$, $r \neq 0$. The solutions for transient flow conditions are beyond the scope of this book and can be found in the cited literature. Figure 3.5 depicts a comparison of calculated and measured data on the transient behavior of the matric potential (h) for surface (Fig. 3.5a) and subsurface (Fig. 3.5b) emitters with a constant flow rate of 1.6 l/h (Coelho and Or 1997). Note the time required for attainment of steady-state flow conditions at different distances from the emitters.

The linearity of these analytical steady-state and transient flow solutions make them amenable to the principle of superposition in space and time. This is

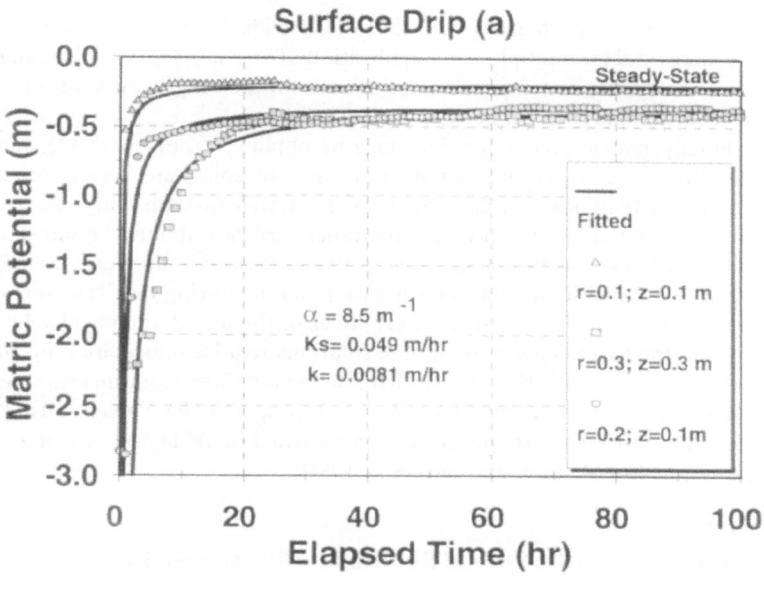

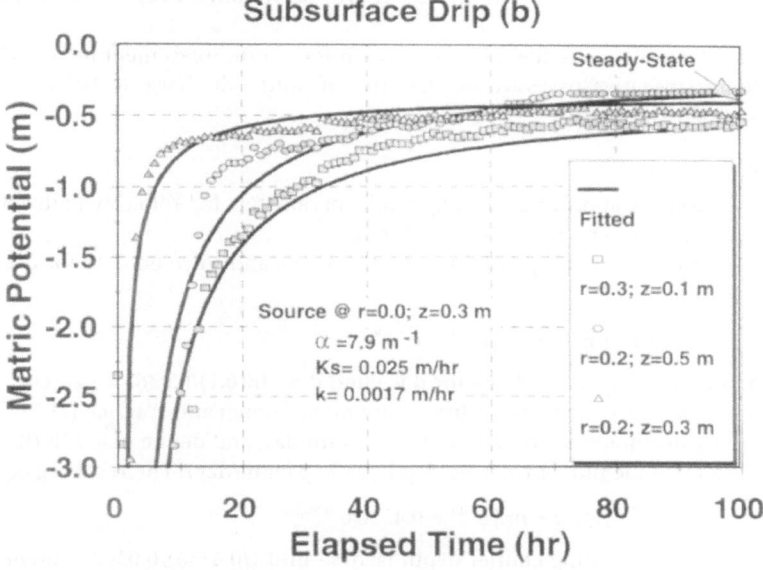

Fig. 3.5. Measured and calculated values of matric potential as a function of time at different locations in the wetted volume for surface and buried sources (z = 0.3 m) with initial water content = 0.18 m³/m³ and flow rate = 1.6 l/h (Coelho and Or 1997)

particularly useful when one considers the distribution of matric potential in an array of sources and for multiple water applications. For example, for cyclic water input (e.g., irrigation cycles), or temporal variations in source strength, the value of ϕ may be calculated by superposition in time (Warrick 1974).

Finally, matric potential values may be obtained from Eq. (3.12), and the corresponding $\phi(r,z,t)$ can be obtained using any soil water retention model (van Genuchten 1980; Russo 1988). Solutions for transient-state flow were very useful in describing actual soil water dynamics under surface and buried point sources in field and laboratory studies (Or and Coelho 1996). Figure 3.6 illustrates the measured and calculated dynamics of soil water matric potential during two irrigation cycles (with no plant root uptake). Illustrative examples of the use of analytical solutions will be given in the following section. These solutions require information on the hydraulic properties of the soil K_s and α which may be obtained from measurements of K vs. h. In many situations, however, the hydraulic properties for a given soil are not known, and values for a similar soil, such as those listed in Table 3.1, may be used as a first approximation (Amoozegar-Fard et al. 1984).

3.1.4
Applications of Analytical Solutions – Illustrative Examples

Example 3.1: Selection of Emitter Depth for Minimal Evaporative Losses

Problem: Estimate the minimum depth for emitter placement in **Yolo clay** soil such that evaporative losses are less than 10% of emitter discharge of 2 l/h (assuming steady state flow).

Solution:
1. Select the appropriate soil hydraulic parameters for Yolo clay (Table 3.1): $\alpha = 3.67 \times 10^{-2}\,cm^{-1}$ and $K_s = 9.33 \times 10^{-6}\,cm/s$.
2. According to Philip (1991a) we take m = ∞, and solve Eq. 3.10 for d assuming E/q = 0.1:

$$E/q = 0.1 = [m/(2+m)]\,e^{-d\alpha} = e^{-0.0367d}$$

3. Rearranging and solving for d we find d = $-\ln(0.1)/0.0367$ = **62.7 cm**
4. We may use the more realistic solution of Lomen and Warrick (1978) by considering maximum evaporation of E_s = 6 mm/day, and define m = $2*E_s/K_s$ = 1.531 (use compatible units by converting K_s to 7.837 mm/day). The resulting equation:

$$E/q = 0.1 = [m/(2+m)]\,e^{-d\alpha} = 0.4336\,e^{-0.0367d}$$

and the resulting emitter depth is: d = $-\ln(0.1/0.4336)/0.0367$ = **40 cm**
5. We repeat the calculations for **Dackley sand** soil (Table 3.1) for the same conditions as above. The calculation according to Philip (1991a) show that the minimum depth is: d = $-\ln(0.1)/0.513$ = **4.5 cm**; and Lomen and Warrick's (1978) m = $2*E_s/K_s$ = 0.143, and the minimum depth: d = $-\ln(0.1/0.143)/0.513$ = **0.7 cm**

6. Two concluding remarks: (a) One should keep in mind that these depths are designed for upper bounds on evaporation assuming steady state conditions. Lomen and Warrick (1978) offer a more realistic solution for transient flow

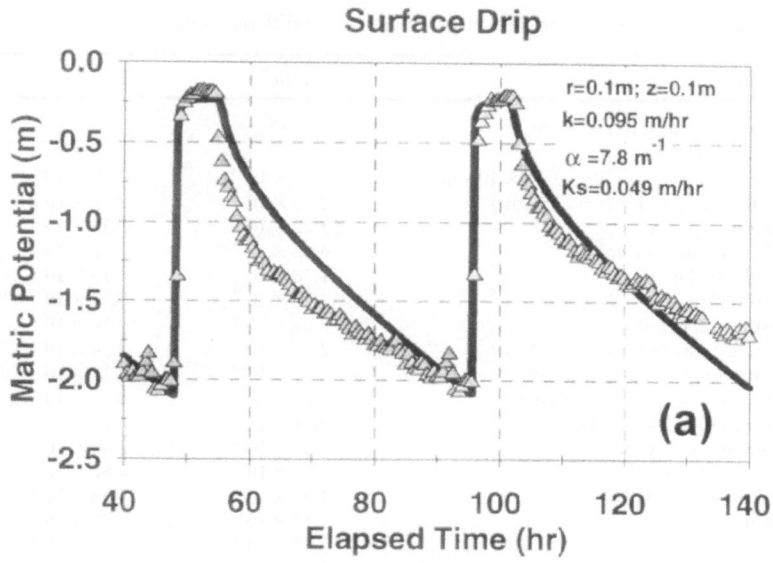

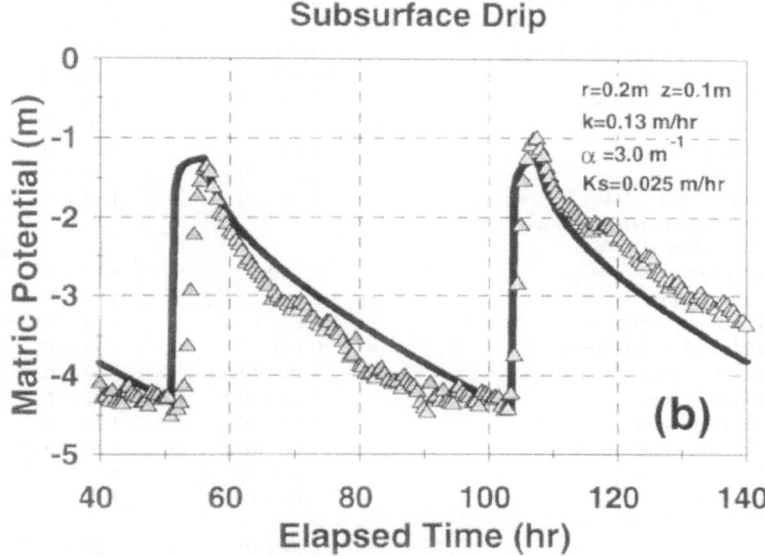

Fig. 3.6. Observed and predicted matric potentials obtained by the analytical solution for transient flow (Warrick 1974) for surface and buried (at 0.3 m depth) point sources at two locations within the wetted volume in Millville silt loam soil. The flow rate was 1.6 l/h, initial water content = 0.18 m³/m³, and measurements were conducted in large containers without plants (Coelho and Or 1997)

Table 3.1. Hydraulic conductivity parameters for different soils (Amoozegar-Fard et al. 1984)

Soil	α (cm^{-1})	K_s (cm/s)	r
For original references see			
Bresler (1978)			
Chino clay ($-1 = h = -1.79 \times 10^4$ cm)	6.85×10^{-4}	2.29×10^{-5}	0.75
Chino clay ($-1 = h = -255$ cm)	2.05×10^{-2}	2.29×10^{-5}	0.92
Lamberg clay	3.27×10^{-1}	3.34×10^{-2}	0.98
Bet Netofa clay soil	6.62×10^{-2}	9.5×10^{-7}	0.99
Lakish clay	1.38×10^{-2}	8.10×10^{-5}	0.99
Yolo clay	3.67×10^{-2}	9.33×10^{-6}	0.99
Sticky clay ($-1 = h = -1 \times 10^4$ cm)	8.64×10^{-4}	2.54×10^{-6}	0.72
Sticky clay ($-1 = h = -170$ cm)	2.91×10^{-2}	2.54×10^{-6}	0.94
Fragmented Lamberg clay	4.10×10^{-2}	2.14×10^{-3}	0.99
Peat ($-1 = h = -1 \times 10^4$ cm)	1.04×10^{-3}	6.13×10^{-5}	0.65
Peat ($-1 = h = -130$ cm)	5.38×10^{-2}	6.13×10^{-5}	0.94
Sheluhot silty clay	7.26×10^{-3}	1.44×10^{-6}	0.95
Touchet silt loam	1.56×10^{-2}	4.86×10^{-4}	0.93
Touchet silt loam	1.03×10^{-1}	6.64×10^{-4}	0.98
Silt Loam	1.39×10^{-2}	5.74×10^{-5}	0.99
Yolo fine sandy loam	2.50×10^{-2}	4.07×10^{-5}	0.79
Plainfield sand fraction (210–250 μ)	2.62×10^{-2}	3.00×10^{-2}	0.21
Plainfield sand fraction (177–210 μ)	0.28	2.00×10^{-2}	0.69
Plainfiled sand (149–177 μ)	0.64	1.40×10^{-2}	0.92
Plainfiled sand (125–149 μ)	0.33	1.06×10^{-2}	0.74
Plainfiled sand (104–125 μ)	0.371	7.30×10^{-3}	0.73
Dackley sand	0.513	1.00×10^{-4}	0.89
Oso flasco fine sand	7.20×10^{-2}	2.00×10^{-2}	0.96
G.E. #2 sand	0.17	1.56×10^{-3}	0.94
Crab creek sand	0.466	1.27×10^{-2}	0.98
G.E. #2 sand	5.75×10^{-3}	2.96×10^{-3}	0.94
Sand (USSL +3445)	6.50×10^{-2}	1.44×10^{-3}	0.97
For original references see			
Warrick et al. (1981)			
Clay loam	0.1258	1.12×10^{-3}	–
Sandy loam	0.1112	1.00×10^{-3}	–
Plainfield sand	0.126	3.44×10^{-3}	0.97
Columbia sandy loam	0.100	1.39×10^{-3}	0.97
Guelph loam	3.40×10^{-2}	3.67×10^{-4}	0.99
Ida silt loam	2.60×10^{-2}	2.92×10^{-5}	0.93
Yolo light clay	1.90×10^{-2}	1.23×10^{-5}	0.94
Gila fine sandy loam	4.43×10^{-2}	2.43×10^{-4}	–
Latene clay loam	3.86×10^{-2}	5.21×10^{-5}	–
Panoche loam	4.16×10^{-2}	1.10×10^{-3}	–
Pima clay loam			
	1.40×10^{-2}	1.15×10^{-4}	–

r = the correlation coefficient; and α and K_s = parameters fitted to the Eq. $K = K_s \exp(\alpha h)$

conditions that would result in somewhat shallower depths or lower actual losses for a given emitter depth; (b) The solutions are valid for emitters represented as point or line source configurations (however, the units of E and q should always be kept compatible, L/T).

Example 3.2: Calculation of Matric Potential Near a Buried Emitter

Problem: The discharge rate from an emitter buried at 25 cm below the surface of **Lakish clay** is 2 l/h. Find the matric potential at a radial distance of 20 cm from the emitter at depths of 10 and at the emitter plane of 25 cm (assume steady state flow).

Solution:
1. Select the appropriate soil hydraulic properties from Table 3.1: α = 1.38 $\times$ 10^{-2} cm^{-1} and K_s = 8.1 $\times$ 10^{-5} cm/s.
2. Calculate $\rho = (r^2 + z^2)^{1/2}$ for 10 cm depth – note that since the origin of the coordinate system is centered on the dripper, a depth of 10 cm below the soil surface is expressed as z = −15 cm (a depth 15 cm above the dripper plane), hence $\rho 1 = (15^2 + 20^2)^{1/2}$ = 25 cm; and for a depth of 25 cm z = 0, and $\rho 2 = (0^2 + 20^2)^{1/2}$ = 20 cm.
3. Next, we employ Eq. 3.8 to calculate the matric flux potential ϕ_B as:
4. $\phi_B(20,-15) = (q/4\pi\rho)\ e^{(\alpha/2)(z-\rho)} = (2000/4\pi25)\ e^{(0.0138/2)(-15-25)}$ = 4.83 cm^2/h
5. We then calculate h(20,−15) from Eq. 3.12 as: h = $1/\alpha\ \ln[\alpha\phi/K_s]$; h(20,−15) = **−107 cm** (note that Ks was converted to 0.292 cm/h).
6. A similar procedure for the second depth of z = 0 cm, yields: ϕ_B = 6.932 cm^2/h, and h(0,20) = **−81 cm.**
7. This procedure may be automated using a computer worksheet to calculate the entire distribution of matric potential and water content around the emitter (assuming steady- state flow conditions).
8. Positive pressures may develop for some distances close to the emitter (Philip 1992) – the criterion for unsaturated conditions based on these linearized solutions is the flow domain where ϕ_B < K_s/α (21.15 cm^2/h in Lakish clay soil)

3.1.5
Analytical Solutions – Design vs. Management Perspectives

Quite often, usefulness of the scientific approach for design is limited by a lack of information on soil hydraulic properties, the lack of consideration of plant root uptake, and the incompatibility of scientifically-based recommendations with commercially available products (e.g., emitter spacing and discharge). The success of empirical design practices propagated by drip manufacturers indicates that the refinements offered by sound scientific considerations of emitter-soil-plant interactions play a secondary role to economic, hydraulic, and availability considerations. Moreover, it is clear that the prevailing tendency is to over-design, thereby enhancing the likelihood for success of a drip system. Realizing that drip systems design is heavily constrained by concerns other than soil-water-plant considerations, the role of analytical solutions such as those based on steady-state flow in design should be more modest. Rather than attempting to answer specific design questions (e.g., a specific emitter spacing), analytical solutions should be used primarily as screening tools for identifying potential ranges of emitter spacing, or for establishing bounds on evaporative losses [e.g., Eq. (3.10)].

A distinction should be made between drip system design considerations and practical management objectives (with their respective information requirements). The management of drip systems requires some understanding of water distribution patterns, either described and predicted analytically, or shown by numerical models. The use of steady-state-based analytical solutions for design purposes considering various source configurations have been proposed by many (e.g., Bresler 1978; Amoozegar-Fard et al. 1984; Warrick 1986; Risse and Chesness 1989). However, design considerations may be quite different than irrigation management considerations. Unlike design, management is a highly dynamic process that need not be conservative. Management guidelines should be flexible and adaptive, capable of correcting design errors and coping with situations not anticipated by design. Consequently, analytical solutions for soil water management under drip irrigation should be capable of considering temporal variations in soil water status and changes in plant root extraction patterns, and thus provide a more realistic representation of field conditions. These requirements are likely to be met by transient flow solutions, as was demonstrated by Or and Coelho (1996).

3.1.6
Models for Water Distribution Based on Simplified Geometry and Volume Balance

For many practical situations, detailed information on matric potential or water content distributions within the wetted soil volume is not necessary, and predictions of the boundaries and shape of the wetted soil volume suffice. Such an approach may also be driven by the lack of information on soil hydraulic properties (required for both numerical and analytical predictions), and when mathematical simplicity is of primary importance. Several simple models for soil-wetting patterns based on volume balance and flow geometry have been developed (Schwartzman and Zur 1986; Ben-Asher et al. 1986; Healy and Warrick 1988). These models provide predictions of wetting front positions as a function of the volume of applied water (emitter discharge × application time), soil porosity, and simple soil intake properties (e.g., saturated hydraulic conductivity K_s, or long-term infiltration rate i_0 – both in dimensions of L/T).

The simplest wetting pattern resulting from water application by a surface point source (or a small pond) into relatively dry soil is hemispherical (with volume given by $V = 2\pi r^3/3$). The relationships between the time-dependent effective radius of the hemispherical wetted soil volume $r(t)$, and the amount of water applied is simply:

$$r(t) = \left[\frac{3qt}{2\pi\Delta\theta} \right]^{1/3},$$

(3.15)

where q is the emitter discharge $[L^3/T]$, t is irrigation time, and $\Delta\theta$ is the average change in volumetric soil water content $[L^3/L^3]$ within the wetted soil volume. This approximation appears to work best for fine-textured soils, dry initial conditions, and relatively short times. The usefulness of this approximation was demonstrated in several studies (as discussed in detail by Ben-Asher et al. 1986). For longer times (and for wetter conditions) the approximation becomes invalid as gravity effects distort the shape of the wetted volume from hemispherical to semi-ellipsoidal.

The elapsed time at which the effect of gravity on the flow process becomes dominant (equals capillarity) and begins to distort the hemispherical approximation is related to the soil properties (and to the change in water content within the wetted soil volume). An estimate for this "gravity time" t_{grav}, for flow from a small pond is given by (Philip 1986):

$$t_{grav} \approx \frac{2\Delta\theta}{\alpha K_s} = \frac{1}{4\alpha K_s}. \tag{3.16}$$

It is assumed that a typical change in volumetric water content (for relatively dry initial conditions) is about $\Delta\theta = 0.125\,\mathrm{cm^3\,cm^{-3}}$ (which represents the difference between water contents at field capacity and wilting point for many soils). The factor of 4 [Eq. (3.16)] may be increased for wetter initial conditions. This estimate of gravity time may be used to predict whether the hemispherical approximation is reasonable for the soil type and the irrigation time t_i. We estimate that for soil types and irrigation times with $t_{grav}/t_i > 10$, the hemispherical geometry is likely to provide reasonable estimates for the dimensions of the wetted soil volume (a large ratio is indicative of a strong capillarity effect, resulting in spherical soil wetting). A similar approximation for spherical wetting from buried emitters (volume of wet soil $V = 4\pi r^3/3$) may be used, with the same criterion for gravity time (t_{grav}).

Schwartzman and Zur (1986) developed a semi-empirical approach to prediction of wetting patterns under surface drip irrigation. They related key parameters affecting water distribution, such as emitter discharge q, saturated hydraulic conductivity K_s, and total water in the wetted volume V, to the vertical depth of wetting z, and lateral diameter of the wetted volume d (measured at its widest point). Based on results presented by Bresler et al. (1971) for a surface dripper (point source) they estimated the following relationships:

$$z = 2.54V^{0.63}\left(\frac{K_s}{q}\right)^{0.45}, \tag{3.17a}$$

$$d = 1.82V^{0.22}\left(\frac{K_s}{q}\right)^{-0.17}, \tag{3.17b}$$

$$d = 1.32z^{0.35}\left(\frac{q}{K_s}\right)^{0.33} \approx 1.32\left(\frac{zq}{K_s}\right)^{1/3}, \tag{3.17c}$$

Consistent units must be used in these approximations such as: d and z [m], q [m³/s], K_s [m/s] and V [m³]. Despite the large degree of empiricism, and the limited data used in their derivation, Eqs. (3.17a–3.17c) offer a simple and useful means for predicting wetting patterns, including the expected distortion in the wetted volume (not predicted by the hemispherical approximation). A few applications of these expressions for emitter spacing design are discussed in Chapter 4. The combined effects of emitter discharge, soil hydraulic properties, and application time on the shape of the wetted soil volume, and the position of the wetting front are illustrated in Fig. 3.7, based on Bresler's (1978) simulation results.

Healy and Warrick (1988) have used numerical solutions of a dimensionless form of Richards' equation for flow from a point source (on the soil surface) to estimate the empirical coefficient of "generalized" equations for the shape of the wetted soil

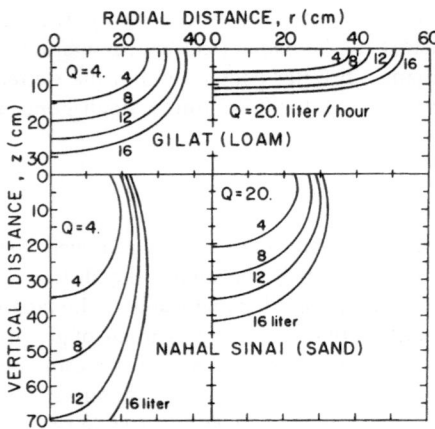

Fig. 3.7. Wetting front position as a function of discharge rate Q and cumulative infiltration in liters (the *numbers labeling the lines*) (Bresler 1978)

volume. The resulting equations and coefficients are applicable for a wide range of soil types and source discharge rates. The general form of the equations is:

$$r_i^* = A_i \tau^{1/2} + B_i \tau + C_i \tau^{3/2}, \tag{3.18}$$

where r_i^* is the dimensionless distance from the source ($r_i^* = \alpha_{vg} r_i$ with r_i the physical distance and α_{vg} a length scaling factor associated with the water retention model of van Genuchten (1980), i is an index of the direction of r: taken vertically (i = 1; r = 0), diagonally (i = 2; r = z), and along the soil surface (i = 3; z = 0), τ is dimensionless time ($\tau = [\alpha_{vg} K_s t]/\Delta\theta$), the coefficients A_i, B_i, and C_i are coefficients dependent on the parameter n (a van Genuchten retention parameter), and dimensionless emitter discharge $q^* = (\alpha_{vg}^2 q/K_s)$. The dimensionless wetted soil volume is also given in an equation similar to Eq. (3.18). Healy and Warrick (1988) presented an extensive table for the coefficients that cover most practical situations (soil types and emitter discharge) and may be useful for design purposes (when soil hydraulic properties are known).

An excellent comparison among various methods for predicting wetting front positions during drip irrigation was presented by Angelakis et al. (1993). These included a finite element solution, Warrick's (1974) analytical solution to the linearized flow equation (Eq. 3.14), the effective hemisphere approximation (Ben-Asher et al. 1986), and the generalized solution of Healy and Warrick (1988). Angelakis et al. (1993) found that all solutions provided better predictions of wetting front positions for clay loam soil than for sandy soil. As expected, the hemispherical model did not capture the effects of gravity distortion on wetting front positions. While diagonal and vertical wetting front positions were estimated reasonably well by the generalized solution of Healy and Warrick (1988), a significant over-prediction of lateral wetting front positions was found. Finally, both the finite element model and the linearized solution provided reasonable predictions of soil water distribution in the wetted volume. A summary of the comparisons conducted by Angelakis et al. (1993) is depicted in Fig. 3.8 for both soil types.

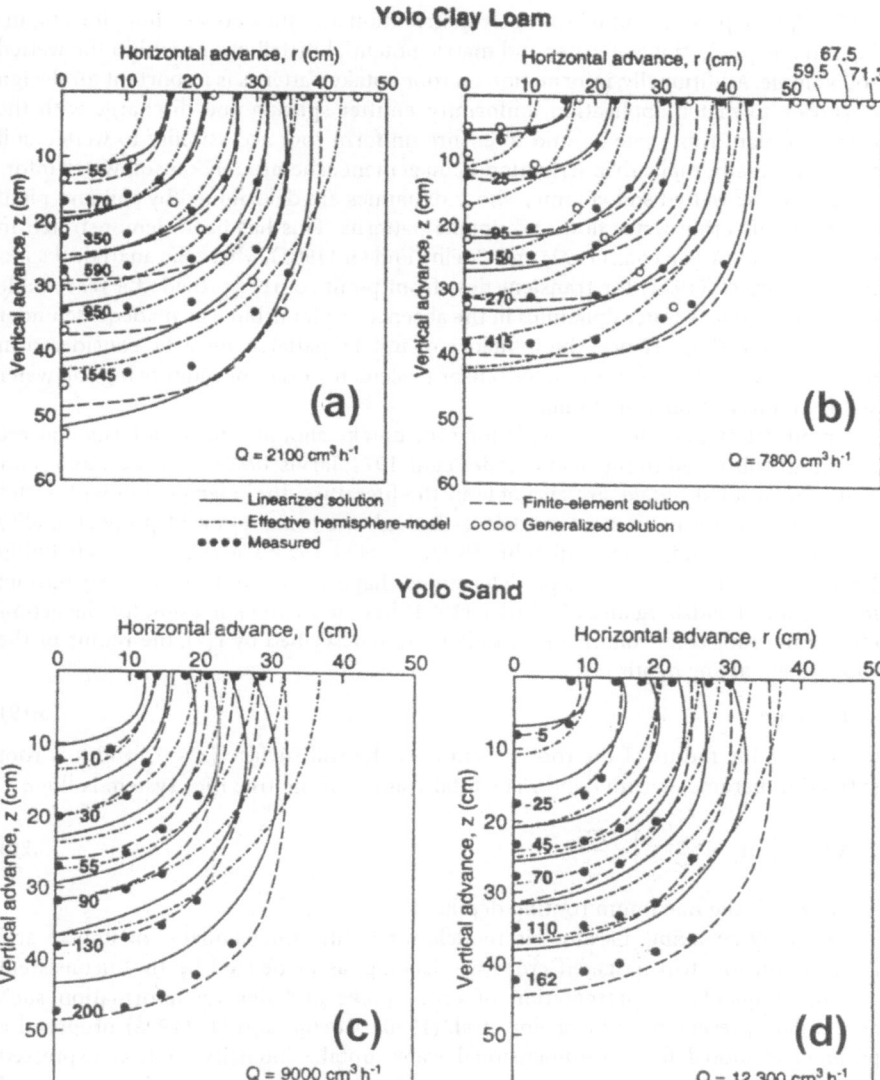

Fig. 3.8. A comparison of methods for estimating wetting front advances under two application rates for two soils at different times (Angelakis et al. 1993)

3.1.7
Plant Water Uptake

Monitoring and modeling soil water distribution for drip irrigation management under cropped conditions requires information on water uptake patterns by plant roots. Uptake patterns influence water distribution and thus are essential for obtaining reliable predictions of water and matric potential distributions within the wetted soil volume. Additionally, information on root uptake patterns is important for design purposes, to match application uniformity, emitter spacing and discharge with the extent of plant root systems, and to ensure uniform root accessibility to wetted soil volumes. Finally, many drip irrigation management schemes rely on soil water information in the wetted soil volume, whose dynamics are determined by soil and plant attributes affecting water flow and uptake patterns. This has been demonstrated in recent studies by Sen et al. (1992) and Coelho and Or (1997), where the analytical solution of Warrick (1974) for transient flow from point sources provided a reasonable description of soil-water dynamics in the absence of plants but was inadequate under cropped conditions. Hence, the influence of uptake patterns must be considered in developing guidelines for soil water sensor placements used for monitoring soil water status and irrigation scheduling.

Empirical or parametric models for root uptake should reflect patterns that are commonly observed in the field (Feddes et al. 1974; Jarvis 1989). Very few models for multidimensional uptake are available in the literature; the interrelation with water distribution patterns in most models is through the assumption of proportionality between uptake and water availability (Neuman et al. 1975; Warrick et al. 1980; Philip 1991b). Others have assumed a predetermined shape for the root density distribution; for example, Landsberg and McMurrie (1984) have used an expression for the geometry of the root zone volume of an isolated tree described by r(z), the radius of the root system at any depth z:

$$r(z) = r_0 e^{-kz}, \tag{3.19}$$

where r_0 is the radius of the root system near the soil surface (z = 0), and k is root extinction parameter with depth. The total volume of the tree root system is then:

$$Vt = \frac{\pi r_0^2}{2k}\left[1 - e^{-2kz\,\text{max}}\right] \approx \frac{\pi r_0^2}{2k}, \tag{3.20}$$

where z_{max} is the maximum rooting depth.

Generally speaking, parametric models for multi-dimensional root uptake and distribution for drip irrigated crops are lacking, as evidenced by the inconsistent and often qualitative presentation of root uptake and density information such as the data presented by Batchelor et al. (1990). Coelho and Or (1996) proposed a parametric model for two-dimensional water uptake intensity (u [r,z] expressed as volume of water extracted per soil volume per unit time) in the wetted volume of drip-irrigated corn. The parametric model is based on bivariate Gaussian distributions for different plant-dripper configurations. The uptake intensity pattern on a plane (representing a soil cross section) reflects the combined effects of distance from the source with the non-uniform soil water distribution and availability.

The domain where uptake occurs was characterized by the position of the dripper relative to the plant row and the presence of no-uptake boundaries defined by: (1) the soil surface; and (2) the borders of the wetted soil volume beyond which water contents are prohibitively low. Figure 3.9 illustrates the four basic plant-dripper configurations and the resulting uptake patterns for corn observed in containers and in the field by Coelho and Or (1996). For a surface dripper on a crop row (Fig. 3.9a),

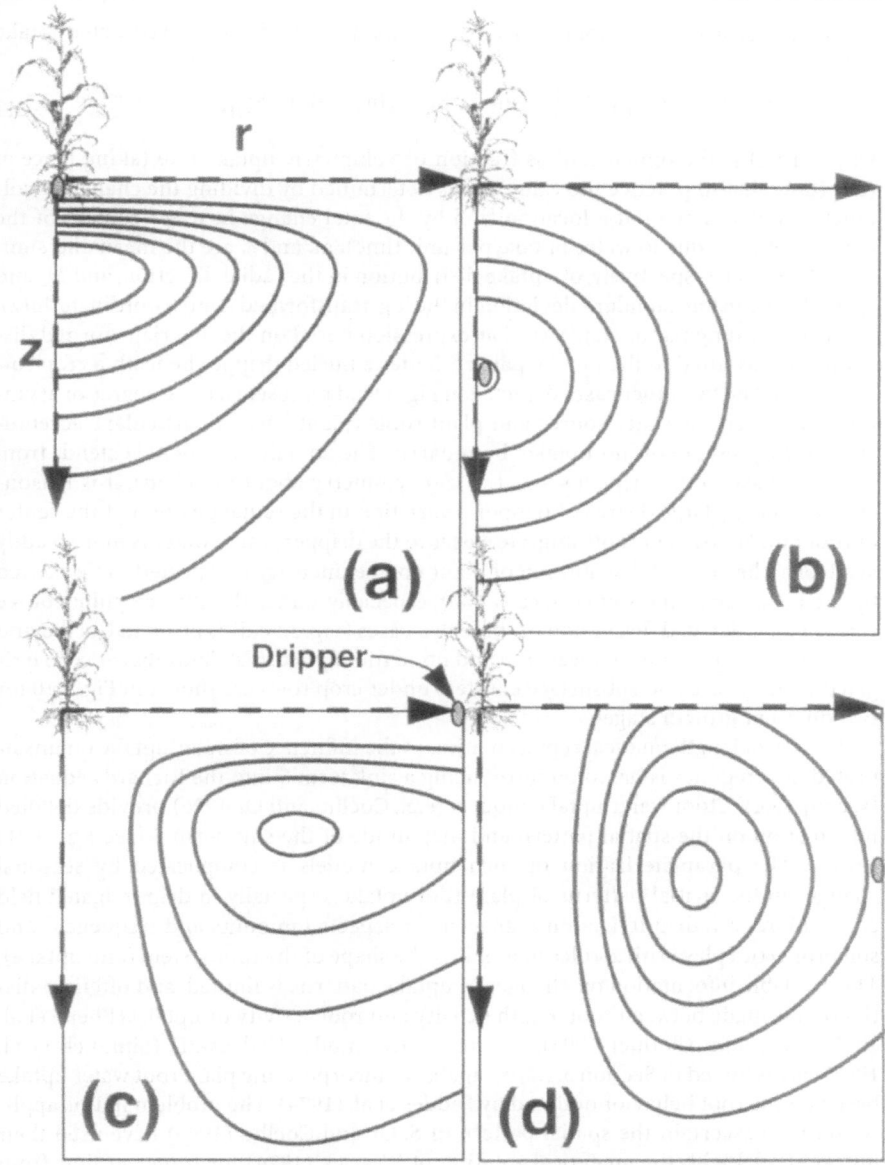

Fig. 3.9. Four configurations commonly found in drip irrigation and the hypothesized uptake intensity patterns. **a** Skewed pattern in the vertical (z) direction and normal in the radial (r); **b** normal in both r and z directions; **c** and **d** skewed uptake in both r and z directions (Coelho and Or 1996)

a Gaussian bivariate semi-lognormal distribution was fitted to observed water uptake patterns:

$$u(r,z) = \alpha/2\pi s_r S_z z \exp\left\{-1/2\left[(r-mr)^2/Sr^2 + (\ln z - M_z)^2/S_z^2\right]\right\}, \qquad (3.21)$$

where $u(r,z)$ is the dimensionless fraction of volumetric uptake rate taking place at (r,z) (note that in practice the value of u is determined by dividing the change in volumetric water content at a location $[r,z]$ by the total change in water content of the entire root zone due to water uptake per unit time), m_r and s_r are the mean and standard deviation, respectively, of uptake distribution in the radial direction, and M_z and S_z are the mean and standard deviation in the log-transformed depth coordinate, $\ln(z)$, and α is a scaling parameter. A similar expression based on the bivariate normal distribution was fitted to the uptake pattern under a buried dripper beneath a crop row (Fig. 3.9b). The two other cases depicted in Fig. 3.9c,d represent a large degree of asymmetry between the water source and plant roots (plant base in particular), accentuated by the presence of "no-uptake" boundaries. The domain considered extends from the plant base to the dripper's axis (i.e., no symmetry about the plant). It is reasonable to expect a large degree of temporal variation in the radial position of the center of root uptake, as plant roots migrate closer to the dripper where water is more readily available. The vertical distribution of plant uptake intensity is expected to be skewed by the presence of the soil surface ($z = 0$), especially under the surface point source (Fig. 3.9c). A detailed discussion of these two cases (Fig. 3.9c,d) is given in Coelho and Or (1996). Comparisons of measured and fitted root uptake distributions for corn irrigated with surface and subsurface emitters under crop rows are shown in Fig. 3.10 for two different growth stages.

The formal mathematical representation of the influence of water uptake on unsaturated flow regimes is based on introducing a sink term S into the Richards equation [see Eq. (3.4)]. Root water uptake models (e.g., Coelho and Or 1996) provide detailed information on the spatial pattern and magnitude of the sink term S [see Eq. (3.21) above]. The parameterization of such uptake models is complicated by seasonal changes in the spatial patterns of plant root uptake, especially in drip irrigated field crops. Moreover, drip irrigation management aspects (amounts and frequency) and soil properties play critical roles in molding the shape of the root system (and uptake). The available information on changes in uptake patterns is limited, and often no distinction is made between root length density and root activity or uptake (Phene et al. 1991; Green and Clothier 1995). The numerical model Hydrus-2D (Simunek et al. 1993), as discussed in Section 3.1.2, is capable of incorporating plant root water uptake based on the root behavior outlined by Feddes et al. (1974). The problem in this application is to ascertain the spatial pattern of S. Or and Coelho (1996) have used their uptake model with the analytical solution of Warrick (1974) for transient flow from point sources to describe soil-water dynamics under drip irrigated corn at selected locations. The idea of modeling soil water dynamics at a few locations in the wetted root zone is appealing, in particular, for the selection of potential locations for soil water sensors (most of which are point measurements which "sense" changes in their immediate neighborhood).

The *localized water balance* approach (Or and Coelho 1996) computes changes in water content at a point (or a small soil volume) as a superposition of two processes, water flow (in or out of that volume), and the fraction of water uptake

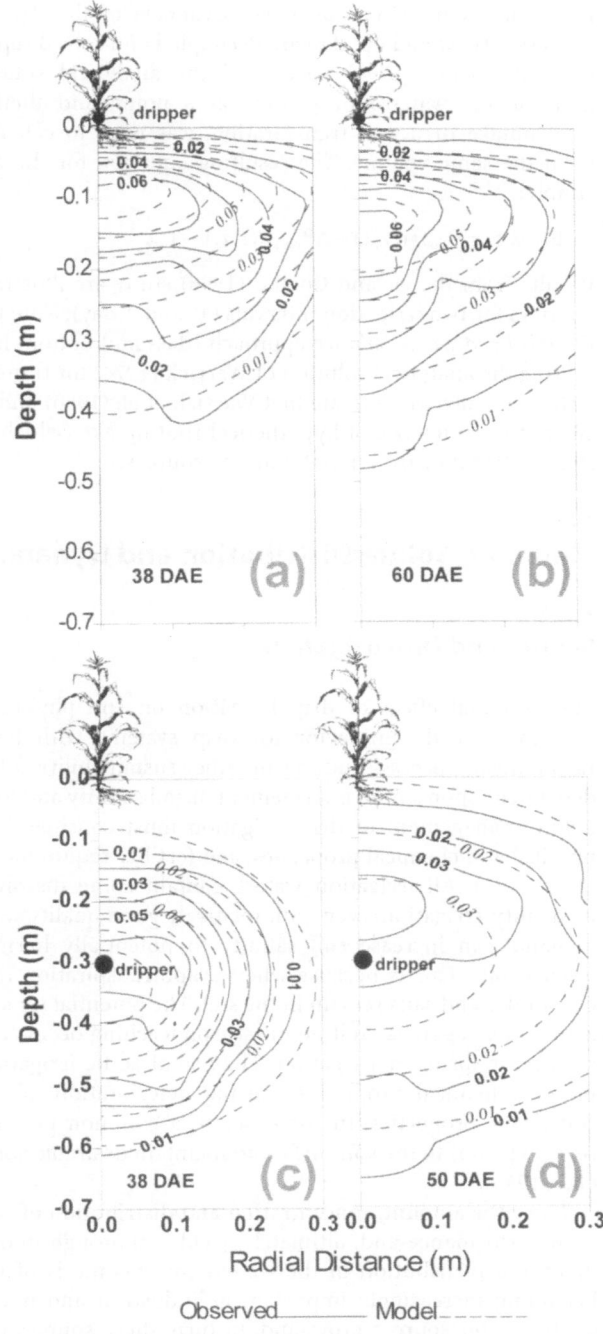

Fig. 3.10. Calculated normalized values using the bivariate semi-lognormal model and observed normalized values of uptake intensity around the surface drip source in a container with 2-day irrigation interval, **a** 38 days after emergence (*DAE*), and **b** 60 DAE; for a subsurface drip source, **c** 38 DAE with a 2-day irrigation interval and **d** with a 1-day interval (Coelho and Or 1996)

from that volume. The primary advantage of this approximation is that the two processes (flow and uptake) are decoupled and solved separately. In their application, Or and Coelho (1996) have used the analytical solution of Warrick (1974) to describe the water flow process at a point, and then Eq. (3.21), multiplied by total uptake or actual transpiration, was used to calculate the fraction of uptake from the same volume. The resulting equation for the changes in water content is simply:

$$\theta(r, z, t) = \theta_{flow}(r, z, t) - \Delta\theta_{uptake}(r, z, t - t_0). \tag{3.22}$$

Results from the Or and Coelho (1996) study are illustrated in Fig. 3.11 for surface emitters for two irrigation intervals (1- and 2-day). Note the large difference between the *localized water balance* approach (data and bottom line), and the top line representing the analytical solution of Warrick (1974) for transient flow with no plant root uptake. Finally, we reiterate that Warrick et al. (1980) and Philip (1991b) have considered the influence of hypothetical root uptake (which varies with depth only) on steady-state flow from point and line sources.

3.2
Modeling Solute Distribution and Dynamics

3.2.1
Solutes and Drip Irrigation

The potential effect of drip irrigation on soil physical and chemical properties is a primary design factor for drip systems with low-quality irrigation water. Economical success and agronomic sustainability of many drip systems are dependent upon sound management of soil salinity and fertility of the crop root zone. Solute management of drip irrigation must consider irrigation water quality, soil physical and chemical properties, and fertility requirements of the crop (Nightingale et al. 1985). All irrigation waters contain some dissolved salts, which determine its quality. Irrigation, even with relatively good quality water under high evaporative demand, can increase soil salinity to potentially harmful levels unless salts are leached out. This is because when evapotranspiration takes place, only pure water evaporates and salts remain in the soil. The potential for salt accumulation in the root zone can be aggravated if no off-season leaching occurs either by precipitation or by surface or sprinkler irrigation. The use of sodic irrigation water may alter the soil physico-chemical properties causing deterioration of soil structure and reduced soil intake properties. In some places, a common practice is to apply a Ca source (e.g., gypsum) to the soil surface to maintain desirable soil intake properties (Mantell et al. 1985).

Since the amount, concentration and distribution of solutes in the soil determine crop performance and, ultimately, yield, a thorough understanding of solute movement and distribution in the wetted soil volume is of utmost importance. This is becoming increasingly important as industrial and municipal demands on good-quality water sources grow, and, in turn, these sources of irrigation water must be replaced by alternative low-quality water sources (e.g., saline water and treated effluent) for drip irrigation (Mantell et al. 1985; Oster 1994). Proper fertigation man-

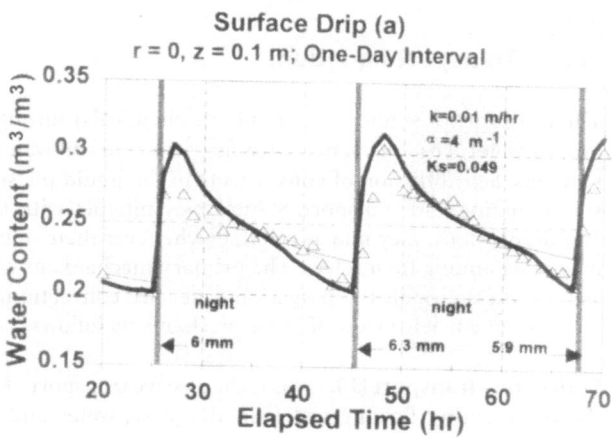

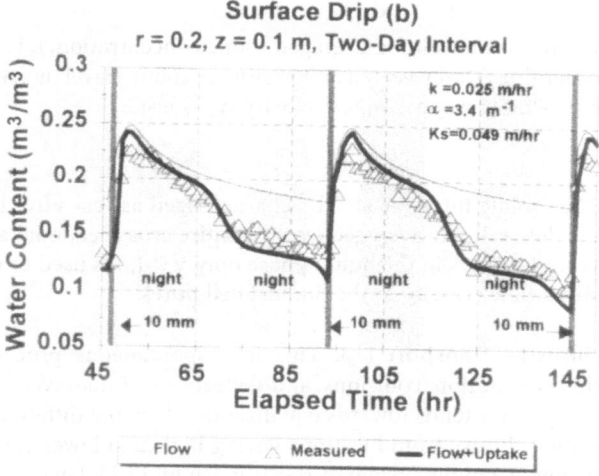

Fig. 3.11. Measured and calculated soil water contents during two irrigation cycles for a radial distance r = 0.0 m, depth z = 0.1 m for a 1-day irrigation interval; and **b** r = 0.2 m, z = 0.1 m for a 2-day interval; for a surface dripper on a crop row with a flow rate of 1.6 l/h. Calculations corrected for plant water uptake using the Warrick (1974) model for transient flow model are denoted by a *thick line* (Or and Coelho 1996)

agement with drip irrigation requires consideration of fertilizer transport properties (mobility, adsorption, etc.), and desired concentration levels within the wetted soil volume. This is particularly important for enhancing crop uptake opportunity for highly mobile fertilizers (e.g., nitrate) and for introducing sufficient amounts of fertilizers of low mobility (e.g., phosphorous) at the same time into the same root zone soil volume.

3.2.2
Solute Transport Equations

The transport of solutes (water-soluble chemicals) through soils is coupled with the flow of water. The convective streams of flowing soil water carry salts and other constituents. Self-diffusion of constituents in the liquid phase is another mechanism for solute mixing and transport. Solutes may interact with the soil matrix (adsorption and desorption), they may precipitate whenever their solubility is exceeded, and they may react among themselves. The primary mechanisms controlling the movement of salts in the soil with the irrigation water are convection, diffusion, and mechanical dispersion; a brief review of these mechanisms follows.

Convective Transport (J_c). This is the passive transport of dissolved constituents with the representing flowing water. In this case, water and solutes move at the same average rate,

$$J_c = J_w \, c = c\left[K(h)\frac{dH}{dz} \right],$$
(3.23)

where c is the volume-averaged solute concentration, J_c is the solute flux, and J_w is the water flux (Darcy velocity). To estimate solute travel or arrival times, the mean apparent velocity or pore water velocity (v) is used:

$$v = \frac{J_w}{\theta}.$$
(3.24)

Thus solute flux may also be characterized as $Jc = v\theta c$. The water flux J_w represents the flow velocity averaged over an entire cross-sectional area. However, because convection occurs in the liquid phase only, v ($>J_w$) is used to represent the average interstitial flow velocity in the liquid-filled pores.

Diffusive Transport (J_d). This is a spontaneous process resulting from random thermal motion, collisions, and deflections of dissolved molecules. The net effect of this process tends towards equalization of spatial differences in concentration, where solutes diffuse from locations having higher to lower concentration. The rate of diffusion (J_d) in bulk water at rest is given by *Fick's Law*:

$$J_d = -D_0 \frac{dc}{dz},$$
(3.25)

where D_0 is the diffusion coefficient in bulk water. The diffusion coefficient in porous media is lower than for bulk water. Because air and solid particles form barriers to liquid diffusion, the apparent soil-liquid diffusivity D_s [L^2/T]) is a function of the available path for diffusion determined by the tortuosity $T(\theta)$, resulting from the geometry of the medium (i.e., texture and structure) and the volumetric water content. An example of the relationship between D_0 (bulk water) and D_s (soil) is given by Jury et al. (1991):

$$D_s = D_0\theta T(\theta) = D_0 \frac{\theta^{10/3}}{n^2},$$
(3.26)

where n is porosity. The flux of diffusing solutes in an unsaturated porous medium is thus:

$$J_d = -D_s \frac{\partial c}{\partial z}. \tag{3.27}$$

Dispersive Transport (J_h). Differences in flow velocities at the pore scale (due to different pore sizes, shapes, and connectivity) cause the solute to be transported locally at different rates and thus lead to mixing (or dispersion) of an incoming solution within an antecedent solution. The process is macroscopically similar to mixing by diffusion (thermal motion); however, it is entirely dependent on water flow (i.e., not driven by concentration gradients). The solute flux due to mechanical (or hydrodynamic) dispersion (J_h) is described by an equation similar to Fick's Law for diffusion:

$$J_h = -D_h \frac{\partial c}{\partial z}, \tag{3.28}$$

where D_h is the hydrodynamic dispersion coefficient $[L^2/T]$. This coefficient is dependent on the interstitial pore water flow velocity v $[L/T]$), and on the dispersivity λ [L] of the soil (a function of pore sizes and shapes) according to:

$$D_h = \lambda \left(\frac{J_w}{\theta} \right)^a = \lambda v^a, \tag{3.29}$$

where a is an empirical factor usually assumed to equal 1 (i.e., a linear dependency of D_h on v). The value of λ may range from 1 cm in small columns to a few meters in field experiments. In most cases the relative effect of hydrodynamic dispersion can exceed that of diffusion. Because of the macroscopic similarity between diffusion and hydrodynamic dispersion, it is common to combine their coefficients (assuming that they are additive) into a *diffusion-dispersion coefficient* (D_e):

$$D_e(\theta, v) = D_s + D_h. \tag{3.30}$$

The Convection-Dispersion Equation (CDE) for Inert and Nonadsorbing Solutes.
The total flux of dissolved solutes in soil (J_s) is the result of combined transport by the three mechanisms discussed above, and may be described by the *convection-dispersion model*:

$$J_s = -D_e \frac{\partial c}{\partial z} + J_w c, \tag{3.31}$$

where J_s is the total mass of solute transported across a unit cross sectional area of soil per unit time, J_w is the water flux (Darcian flux), D_e is the combined diffusion-dispersion coefficient, and $\partial c/\partial z$ is the spatial solute gradient (partial derivative indicates that the gradient may also vary with time). D_e is dominated by the dispersion process under most flow conditions. Combining Eq. (3.33) with the continuity equation (conservation of solute mass) yields:

$$\frac{\partial(\theta c)}{\partial t} = -\frac{\partial J_s}{\partial z}, \tag{3.32}$$

where θc is the mass of solutes in solution. The continuity equation may be written as:

$$\frac{\partial(\theta c)}{\partial t} = -\frac{\partial}{\partial z} \left(J_w c - D_e \frac{\partial c}{\partial z} \right). \tag{3.33}$$

Assuming steady-state water flow in a homogeneous soil profile (J_w and θ are constant in time and space), reduces Eq. (3.33) to the familiar form of the *convection-dispersion equation* (CDE) for inert and non-adsorbing solutes:

$$\frac{\partial c}{\partial t} = D\frac{\partial^2 c}{\partial z^2} - v\frac{\partial c}{\partial z},$$ (3.34)

where $D = De/\theta$, and $v = J_w/\theta$. The CDE may be expanded to describe two- and three-dimensional solute transport:

$$\frac{\partial c}{\partial t} = \nabla \cdot (D\nabla c) - v \cdot \nabla c.$$ (3.35)

These forms of the CDE become more complicated when modeling reactive and adsorbing solutes. In general, additional terms to account for various forms of solute adsorption and transformations may be incorporated into the same basic form of Eqs. (3.34) and (3.35).

3.2.3
Numerical Solutions for Solute Transport under Drip Irrigation

It should be emphasized that the modeling of solute transport in soils is linked with water flow; hence a solution of the water flow problem must precede the solute transport problem. A finite difference numerical model for simultaneous transport of water and non-reactive solutes under drip irrigation was developed by Bresler (1975), and tested by Bresler and Russo (1975). Subsequent work by Bresler and Green (1987) expanded this model to describe transport of degradable solutes (such as pesticides) under drip irrigation. There are a few, more modern, numerical codes for solute transport from point sources, including the Hydrus-2D model (Simunek et al. 1993, discussed in previous sections on water flow). The Hydrus-2D with its user-friendly graphical interface (and with a bit of training) renders modeling solute transport under drip irrigation a relatively simple task.

The details of Bresler's (1975) procedure for solving salt concentration fields are beyond the scope of this book; however, some of the results from Bresler and Russo (1975) will be presented and discussed. Figure 3.12 shows the results of computations of salt-concentration fields for two soils with different hydraulic conditions, but with the same initial salt concentration in the soil volume (11.25 meq/l), the same amount of total solution infiltrated (12 l), and the same salt concentration at the soil inlet (1.34 meq/l). The last is obtained as the product of the salt concentration in the irrigation water (3.0 and 5.0 meq/l for Gilat and Nahal Sinai soil, respectively) and the water content at saturation. These identities make it possible to compare the salt distribution in the two soils with two drip discharge rates (Fig. 3.12). The plots show an initially saline soil leached by drip irrigation with water of good quality. The leached part of the soil is deeper and narrower in the sandy soil than in the loamy soil. The lower and higher discharge rates have the same effect. The leached salts tend to accumulate close to the wetting front in the sandy soil and are dispersed over a greater volume in the loamy soil. The shape of the leached volume is very much affected by the emitter discharge rate. This may be of importance in bringing saline soils under cultivation by leaching salts from the root zone.

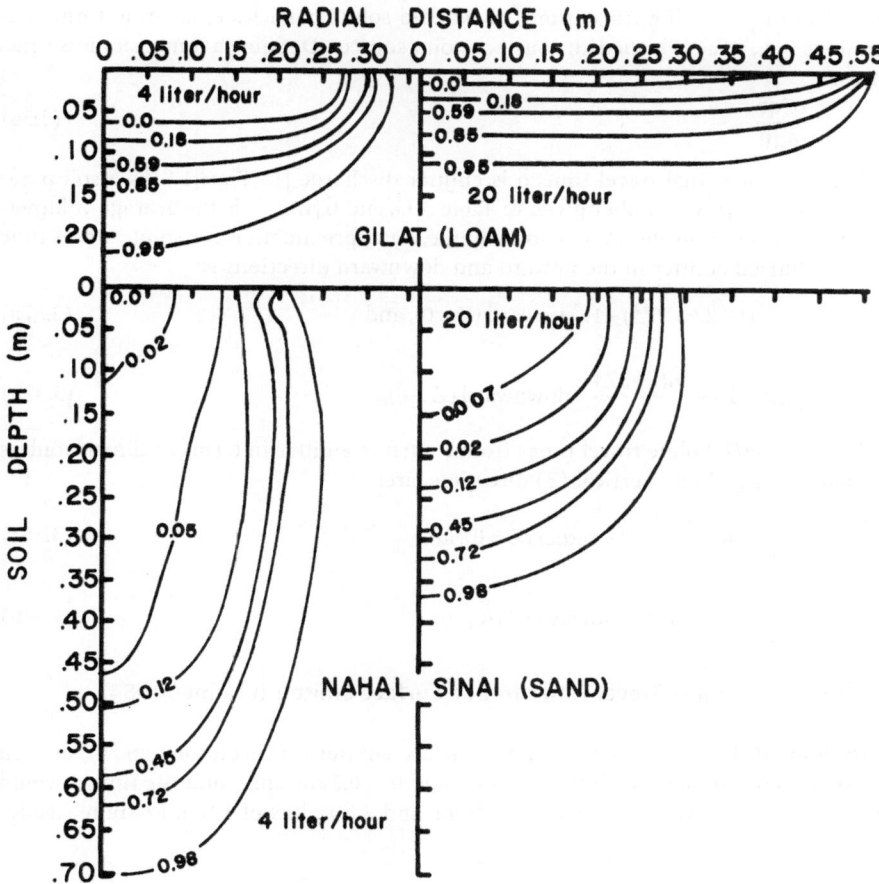

Fig. 3.12. Computed salt concentration fields for two different trickle discharges for two soils. The numbers labeling the curves indicate relative concentrations of salt ($C = C_o/C_n$) after 12 l of irrigation water infiltration. C_o = irrigation water concentration and C_n = initial soil solution concentration (Bresler 1975)

3.2.4
Analytical Solutions for Solute Transport

Most available analytical solutions to solute transport in unsaturated soils described by the CDE are for one-dimensional transport only (Biggar and Nielsen 1967; van Genuchten and Wierenga 1986). A few analytical solutions exist for three-dimensional solute transport under unsaturated, one-dimensional steady water flow conditions (Leij et al. 1991b; Ellsworth and Butters, 1993). To date, no analytical solutions for the general case of simultaneous water flow and solute transport under drip irrigation have been developed. There are, however, a few useful analytical tools based on the extension of solutions to the steady water flow problem by considering purely convective transport of solutes (Clothier 1984; Philip 1984). Philip (1984) considered the

problem of predicting travel times of marked solute particles emanating from continuous (i.e., steady) buried and surface point sources. Defining a dimensionless time,
T as:

$$T = \frac{\alpha^3 qt}{16\pi\theta}, \tag{3.36}$$

where t is the actual travel time, q is emitter discharge $[L^3/T]$, $\alpha[L^{-1}]$ is a soil parameter (the sorptive number given in Table 3.1), and $\bar{\theta}\,[L^3\,L^{-3}]$ is the average volumetric water content in the wetted soil volume, an approximation for solute travel time
from a **buried** emitter in the upward and downward directions is:

$$T = \frac{1}{2}[e^{2Z}(1 - 2Z + 2Z^2) - 1] \quad \text{upward } (Z < 0), \text{ and} \tag{3.37a}$$

$$T = \frac{1}{2}(Z^2 - Z) + \frac{\ln(1 + 2Z)}{4} \quad \text{downward } (Z > 0), \tag{3.37b}$$

where $Z = \alpha z/2$. Solute travel times from a **surface** emitter in terms of dimensionless
radial ($R = \alpha r/2$) and vertical (Z) directions are:

$$T = 2e^R\left(1 - R + \frac{R^2}{2}\right) - 2 \quad \text{radial } (Z = 0), \text{ and} \tag{3.38a}$$

$$T = \frac{Z^2}{2} - Z + \ln(1 + Z) \quad \text{downward } (R = 0). \tag{3.38b}$$

Example 3.3: Solute Travel Times from a Surface Emitter (Clothier, 1984)

Problem: If the discharge rate from a surface emitter is 0.36 l/h ($360\,\text{cm}^3$/h), the soil
is **Manawatu fine sand** with $\alpha = 0.3\,\text{cm}^{-1}$, and $\bar{\theta} = 0.2\,\text{cm}^3\,\text{cm}^{-3}$, find the time it would
take nitrate to travel to 1 a depth of 30 cm, and 2 a radius of 10 cm (assume steady-
state flow).

Solution:
1. Calculate $Z = \alpha z/2$ and $R = \alpha r/2$, $Z = 4.5$ and $R = 1.5$.
2. Find T(Z) and T(R) using eq. (3.38):

    (a) $T(Z) = 4.5^2/2 - 4.5 + \ln(1 + 4.5) = 7.33$.

    (b) $T(R) = 2e^{1.5}(1 - 1.5 + 1.5^2/2) - 2 = 3.60$.

3. Convert T to t using eq. (3.36):

    (i) $t(z) = 16\pi\bar{\theta}T(Z)/(\alpha^3 q) = 7.58$ h

    (ii) $t(r) = 16\pi\bar{\theta}T(R)/(\alpha^3 q) = 3.72$ h

4. The results of these calculations were compared by Clothier et al. (1984) with
   measurements and are depicted in Fig. 3.13. Inspection of Fig. 3.13 shows that
   these predictions are in reasonable agreement with measured values for Bromide
   fronts.
5. The most problematic parameter in these approximations is obtaining a reliable
   value for the average water content in the wetted volume $\bar{\theta}$.

Fig.3.13. Predicted isochrones (*solid curves*) during steady three-dimensional flow in Manawatu fine sand. Bromide fronts (*dashed curves*) at the indicated times are also shown. The cavity radius is $r_0 = 0.004$ m (Clothier 1984)

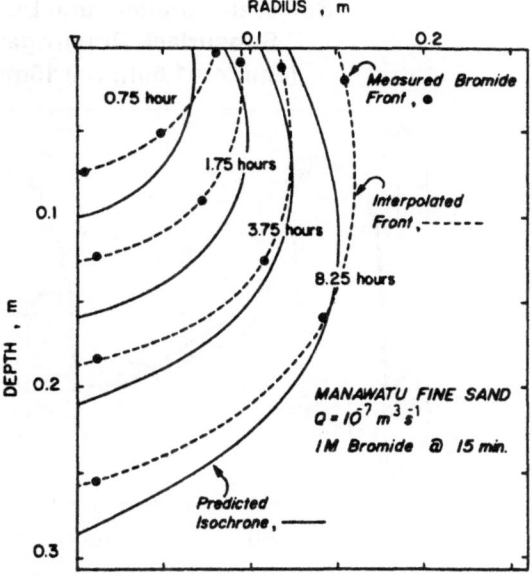

Summarizing, solute travel time estimates obtained by these equations should be viewed as upper bounds (if there are no preferential flow pathways such as cracks). It is relatively simple to recast Eqs. (3.37) and (3.38) to calculate travel distances as a function of time since solute application.

3.2.5
Solute-Plant Interactions

The objective of drip irrigation solute management is to enhance crop growth conditions by reducing amounts of deleterious solute accumulation, increasing availability of beneficial nutrients, and avoiding contamination of groundwater resources. These processes are taking place within a very limited soil volume (along with water uptake) where solute dynamics are rapid as manifested by large temporal fluctuations in solute concentrations, or short residence time for excess irrigation of mobile solutes. An example of the extent of fluctuations in soil water content and solute concentration is depicted in Fig. 3.14 measured in corn grown in silt loam soil and irrigated with a 4 l/h surface emitter (Mmolawa 1999). The effects of solute concentration on crop response and yield result mainly from decreasing the osmotic component of soil water potential and hence limiting water availability to the crop. Toxic effects that are specific to the crop and the specific salt also need to be considered (e.g., boron). These may be toxic to various plant physiological processes, or cause nutritional disorders. A detailed list of crop tolerance and sensitivity levels to various salts is given in Bresler et al. (1982).

The maintenance of tolerable salinity levels in the limited root zone volume under

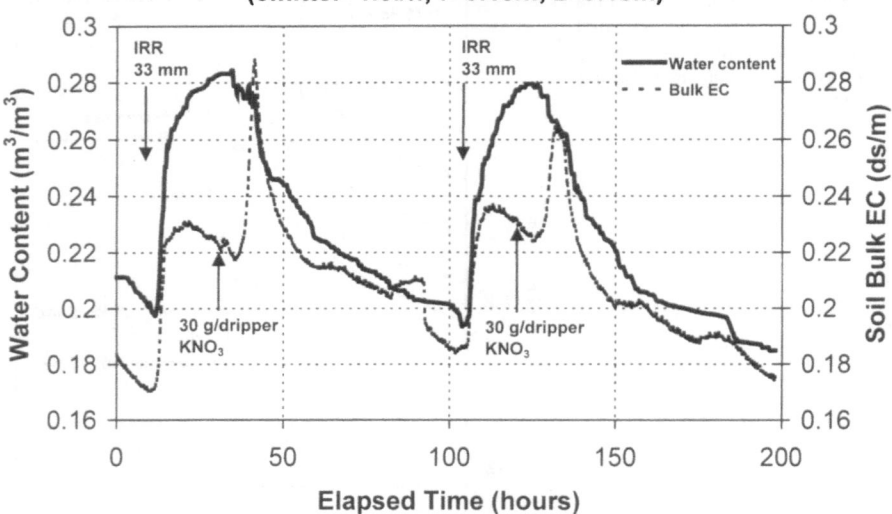

Fig. 3.14. Soil water content and EC dynamics in subsurface drip-irrigated corn (Mmolawa 1999)

drip irrigation with low-quality water (i.e., water with a high content of dissolved salts) is dependent on: the water quality, leaching fraction, irrigation frequency, crop uptake, soil properties and drip system design. An example of the effect of leaching fraction on salt distribution from a drip line source is given in Fig. 3.15. Note the lower salt concentration distribution for the higher leaching fraction (LF) defined as: "the actual fraction of applied water that passes through the plant root zone, also normally expressed in terms of equivalent surface depth" (Bresler et al. 1982).

$$LF = \frac{D}{I}, \tag{3.39}$$

where I and D are irrigation and drainage equivalent depths, respectively.

Many studies have shown that the combined effects of solute convection by water flow and selective water uptake by crop roots (leaving salts behind) result in relatively low salt concentrations near the emitter and increased concentrations toward the fringes of the wetted soil volume (Mantell et al. 1985; Ayars et al. 1985). Studies have shown that high irrigation frequency is more effective in maintaining lower salt concentrations in the proximity of the emitter (or active root zone) than low frequency irrigation (Nightingale et al. 1985). Ayars et al. (1985) found that a higher irrigation frequency (daily irrigation frequency) resulted in lower average salinity profiles compared with an irrigation frequency of 3 to 4 days. They also measured lower osmotic potentials (determined with thermocouple psychrometers) in the saline plots. The effect of water quality on root distribution was investigated by Mantell et al. (1985).

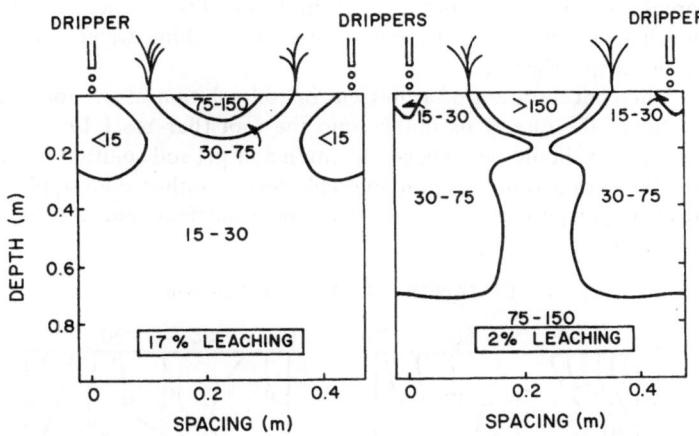

Fig. 3.15. Salt distribution around trickle lines with different leaching fractions. Numbers within each zone represent chloride concentrations in the soil water (mol/m³) (Hoffman et al. 1980)

Figure 3.16 depicts root distribution of drip irrigated cotton with two levels of water quality (measured by their electrical conductivity as 1.0 dS/m and 7.3 dS/m). The crop was irrigated every 3–4 days, and the crop row was 50 cm from the emitter. The results show that the volume occupied by the roots was larger and closer to the emitter for the 1.0 dS/m irrigation water, whereas for lower quality water (7.3 dS/m), plant roots remained restricted to a smaller soil volume beneath the crop row.

One of the distinct advantages of drip irrigation is the ability to accurately control both water and nutrients in the plant root zone. In considering solute-plant aspects of fertigation design and management, Bar-Yosef (1977) listed the following as essential information: 1 the plant daily uptake of various nutrients during the growth period; 2 the relationships between uptake rates and nutrient concentration in the soil solution; and 3 the plant daily water requirement. Table 3.2 (taken from Bar-Yosef 1977) shows the magnitudes of the three components related to N-uptake for drip irrigated tomatoes grown in sand. Bar-Yosef (1977) noted that only 30 to 50% of applied N was taken up by plants; this was attributed to leaching losses and lower

Table 3.2. N-uptake rates, N-concentration in soil solution, and calculated daily water consumption for drip-irrigated tomatoes during various growth stages (Bar-Yosef 1977)

Days after seeding	Daily N uptake (mg/plant/day)	N concentration in soil solution (ppm)	Daily water consumption (l /plant/day)
42–64	65	70 (100)[a]	0.93 (0.65)[a]
64–76	90	106 (140)	0.85 (0.64)
76–111	65	170 (140)	0.48 (0.47)
111–180	98	184 (100)	0.57 (1.05)

[a]The values in parentheses were obtained from direct soil sampling

uptake due to low N-concentrations in the plant root zone. These two potential "loss" mechanisms highlight the importance of matching water and solute applications under drip irrigation.

The concentration of a nutrient in soil solution at the root surface is the main factor of the rate of its uptake into the root (Bar-Yosef 1977; Jungk 1996). Hence, fertigation should be managed to attain a target soil solution nutrient concentration in the wetted root zone (at tolerable levels) rather than apply the same amount over longer periods. The contact between nutrients and the roots can be attained

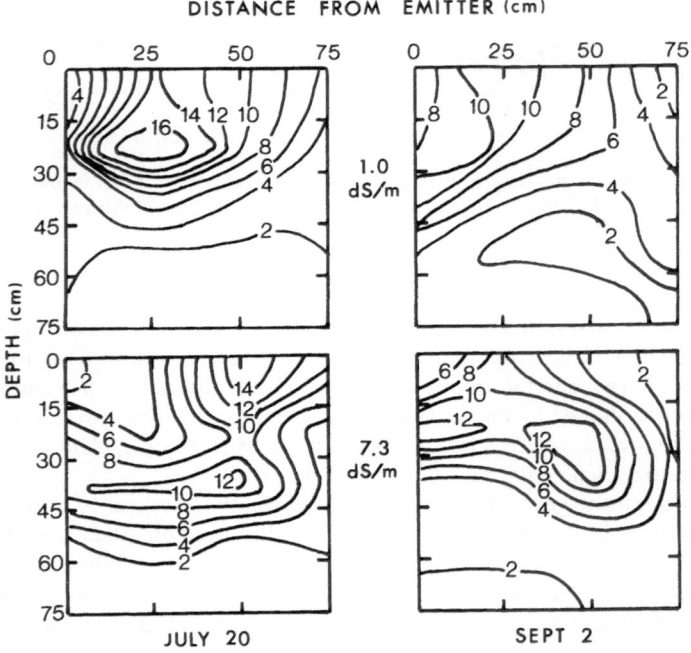

Fig. 3.16. The effect of irrigation water quality on root distribution (% of total weight) (Mantell et al. 1985)

Table 3.3. The relative importance of root interception, mass flow, and diffusion in supplying maize with nutrients (Adapted from Jungk 1996)

Nutrient	Amount needed for 9500 kg/ha grain yield	Approximate amounts supplied by		
		Root interception	Mass flow	Diffusion
		Kg / ha		
Nitrogen	190	2	150	38
Phosphorus	40	1	2	37
Potassium	195	4	35	156
Calcium	40	60	150	0
Magnesium	45	15	100	0
Sulfur	22	1	65	0

either by root growth into soil volumes with nutrients, or by transport of nutrients to roots by mass flow (convective transport) or diffusion (Jungk 1996). The relative importance of the various root-nutrient uptake mechanisms is summarized in Table 3.3. Note the role of mass flow in supplying N vs. the importance of diffusion in supplying P.

Chase (1985) found that vegetable yield was significantly higher when phosphorus was applied by subsurface drip irrigation, than for broadcast application of the same amounts. Though adsorption of P resulted in immobilization near the emitter, extractable P levels and the travel distance from the emitter increased as the application rate increased. It appears that crop response to enhanced P concentrations is partially due to elevated solubility induced by higher localized concentrations and enhanced diffusion in the limited volume, both of which affect the root surface P concentration, and, hence, P uptake.

Drip System Design

4.1
General Considerations

The primary objective of a drip system design is to choose appropriate components and layout to attain adequate distribution of water (and fertilizer) throughout the field to meet crop needs with consideration of economical, operational, water quantity, and water quality constraints. Well-designed drip systems should provide equal (and adequate) soil water availability to all plants in the field at high irrigation efficiency.

Present-day drip irrigation design practices tend to emphasize system hydraulic performance (pressure distribution, filtration, emitter uniformity etc. – all of which may be optimized with the aid of computers), while agronomic-hydrologic considerations, such as emitter-soil-plant interactions, receive less attention, or are dealt with empirically. This disparity may be attributed largely to the complexity of agronomic-hydraulic processes that are usually not amenable to simple design rules. Consequently, there is a tendency to over-design drip systems to ensure successful performance, regardless of detailed agronomic considerations. Though this practice may be adequate for many "standard" scenarios, there is concern that loss of insight may lead to design inflexibility and increased burden on drip irrigation management. Additional concerns are potential escalation of system "over-design" propagated by uninformed drip systems distributors, and erosion of the scientific basis for drip irrigation design and management. The main challenge is to realistically reconcile the complex, and often difficult-to-quantify agronomic-hydrologic aspects with the more straightforward hydraulic aspects, with due consideration of components availability and other constraints.

A cursory inspection of drip systems catalogs and of recent scientific evaluations (e.g. Table 2.1, Hanson 1995) reveals that present standards of uniformity and quality for most emitters are very high. Modern production techniques and advanced product quality control ensure unprecedented application uniformities. Advances in computers and software (Bralts et al. 1995) render many of the tedious hydraulic calculations for routine design and the need for design nomographs, impractical. The value of these tools is primarily for spot-checking computer-based designs; hence only primary principles need to be emphasized.

Variability in plant available water is determined not only by emitter application uniformity, but also by considerations of available soil water variations due to variable soil properties, and the interplay between plant root zone extent and emitter spacing.

4.2
Emitter Spacing and Discharge

For certain vegetable crops, flower beds, or other densely-planted crops with very limited lateral root zones, wetting of the entire area is needed by overlapping the soil volumes wetted by each emitter. Usualy, with row crops the aim is to uniformly wet the plant row, while between the rows the soil may be left dry. In orchards, especially with young trees, a dry zone may separate each tree from its neighbor. Thus, the distances between the emitters along and between the laterals must be adapted to crop water requirements. In addition, these distances should be based on the hydraulic properties of the soil and the discharge rate of the emitters. The attainment of target sizes (volumes) and shapes of wetted soil volumes in relation to peak crop water requirements, irrigation frequency, and crop root zone dimensions may be considered as a unifying design objective for different emitter layouts, spacing and discharge. The procedures for determining emitter discharge and spacing are based on the principles of water flow in porous media discussed in Chapter 3. To keep design considerations simple, we neglect plant water uptake during the infiltration phase of the irrigation cycle.

4.2.1
Emitter Spacing for Non-Overlapping Wetted Volumes

A large number of drip irrigation systems are designed to supply water to non-overlapping wetted soil volumes. Bresler's (1978) approach for emitter spacing design uses analytical solutions for steady-state flow (Wooding 1968) by combining hydrologic information based on emitter discharge and soil properties with desirable soil matric potential values, which, in turn, are based on crop response at a critical location in the crop root zone. Though the methodology is elegant and nicely captures the essential relationships among the various variables, several drawbacks limit its applicability to theoretical studies only (Russo 1993). The primary limitations to practical application of the methodology are the arbitrary choice of the crop response measure (the critical matric potential, hc), and its location (midway between emitters at the soil surface). These parameters were chosen with little consideration of realistic soil water dynamics, i.e., in most cases the surface at the midpoint between emitters is dry, and essentially no plant roots exist at this point. Moreover, the use of such an unrealistic agronomic criterion coupled with an assumed steady-state flow field would lead to over-design of emitter spacing (and discharge) under most practical situations.

The routine use of numerical models (Hydrus-2D), or analytical transient flow models (Warrick 1974) for design purposes is still considered cumbersome and impractical due to the lack of information on soil hydraulic properties, lack of training, and ill-defined hydrologic objectives (what constitutes a "desired" wetting pattern for a crop?). We thus resort to semi-empirical models for the geometry and volume of wetted soil from each emitter, based on the simplest possible inputs (Schwartzman and Zur 1986; Zur 1996).

The empirical coefficients proposed by Schwartzman and Zur (1986) relate the width and depth of the wetted soil volume to emitter discharge and saturated hydraulic conductivity of the soil. The maximal diameter of the wetted volume d is given by (see Eq. 3.17c):

$$d = 1.32 z^{0.35} \left(\frac{q}{K_s} \right)^{0.33} \approx 1.32 \left(\frac{zq}{K_s} \right)^{1/3} , \tag{4.1}$$

where z is the depth of wetting, q is the emitter discharge and K_s is the saturated hydraulic conductivity. The value of d may be used directly to determine emitter spacing for a given wetting depth, discharge and soil hydraulic conductivity, or it may select combinations of d-q for fixed values of the other variables.

Example 4.1: Emitter Spacing-Discharge Combinations for Different Rooting Depths

Problem: Determine emitter spacing-discharge combinations for non-overlapping wetted soil volumes considering various rooting depths in sandy and loamy soils.

Solution:
1. Reasonable values for the saturated hydraulic conductivities are: $K_s = 10$ and 1 cm/h for the sandy and loamy soils respectively (cf. Table 3.1).
2. Select a reasonable range of emitter discharge rates and rooting depths and calculate d using Eq. (4.1) (see Table 4.1).

Zur (1996) expanded the use of these coefficients by introducing the wetted soil volume as a design objective. The basic idea is the selection of the wetted soil volume needed for sustaining the crop's seasonal peak water use (PWU,

Table 4.1a. Emitter spacing (cm) for loamy soil with $K_s = 1$ cm/h

Z_{root} (cm)	Q (l/h)				
	1	2	4	8	12
30	41.0	51.7	65.1	82.0	93.9
60	51.7	65.1	82.0	103.4	118.3
90	59.2	74.5	93.9	118.3	135.4
120	65.1	82.0	103.4	130.2	149.1

Table 4.1b. Emitter spacing (cm) for sandy soil with $K_s = 10$ cm/h

Z_{root} (cm)	Q (l/h)				
	1	2	4	8	12
30	19.0	24.0	30.2	38.1	43.6
60	24.0	30.2	38.1	48.0	54.9
90	27.5	34.6	43.6	54.9	62.9
120	30.2	38.1	48.0	60.4	69.2

cm/day) between consecutive irrigations. This volume is calculated for a given preferred irrigation interval, PI [day], soil water holding capacity WHC (i.e., the difference between field capacity θ_{fc}, and wilting point θ_{wp}), and management allowed deficit, MAD (expressed as a decimal fraction). The wetted soil volume V_w is given by:

$$V_w = \frac{PWU \cdot PI}{WHC \cdot MAD} \cdot DL \cdot d = V_w^* \cdot d, \qquad (4.2)$$

where DL is the spacing between drip lines, d is emitter spacing, and V_w^* lumps all the parameters required for V_w calculation except d (for later calculations). Note that the expression given by Zur (1996) for V_w does not consider the area irrigated by an emitter $DL \cdot d$, and instead, assumes a soil surface area of $1\,m^2$ in Eq. (4.2). Zur (1996) represents the geometry of the wetted volume by a truncated ellipsoid whose center is a distance h below the soil surface. For simplicity, and because the dimensions of h in Zur (1996) are not defined, we shall assume that the wetted volume is approximated by a semi-ellipsoid whose volume is given by:

$$V = \frac{\pi z d^2}{6}, \qquad (4.3)$$

which is equivalent to Zur's (1996) formulation with h = 0. Some of the coefficients in Eq. (4.2) may be approximated for many soils and crops such as: WHC = $0.15\,cm^3\,cm^{-3}$, the value of MAD = 0.3 is a reasonable target value for drip irrigation, and PWU = 0.5–0.7 cm/day in many irrigated areas. These simplifications lead to a simplified form of Eq. (4.2) as: $V_w \approx 13.5 \cdot PI \cdot DL \cdot d\,[cm^3]$. Combining Eqs. (4.1) to (4.3) yields an expression for emitter discharge as a function of soil Ks, and wetted volume shape (d and z):

$$q = \frac{6K_s d V_w}{2.3 z \pi} = \frac{0.83 K_s V_w^* d^2}{z^2}, \qquad (4.4)$$

where $V_w^* = 13.5.\,PI \cdot DL$. The inverse procedure proposed by Zur (1996) proceeds as follows:

1. For a given row spacing, or drip line spacing DL, and preferred irrigation interval PI, compute V_w^*. *(Example: DL = 100 cm, and PI = 3 days, yields V_w^* = 4050 cm²).*
2. Select d values for which to compute the wetted volume V_w, and several possible z values, i.e., apply Eq. (4.2) and then solve Eq. (4.3) for z. *(Example: assuming d = 100 cm, V_w = 405 000 cm³, and z = 6V/πd² = 77.3 cm)*
3. Finally, select the q (Eq. 4.4) that best satisfies the assumed shape and the required wetted soil volume *(Example: assume K_s = 1 cm/h, q = 0.83*1*4050*(100/77.3)² = 5626 cm³/h)*

The implicit assumption behind Zur's (1996) procedure is that for reasonable combinations of z and d, and for a given wetted volume, the rate of water application q (in relation to the soil hydraulic conductivity) may be chosen to attain the desired dimensions and shape of the wetted soil volume. The chosen ellipsoid aspect ratio d/z determines emitter discharge for a given soil hydraulic conductivity (i.e., a larger aspect ratio requires a larger emitter discharge). If the effect of gravity on soil water movement is small, the wetted volume can be approximated as a hemisphere (see Section 3.1.6).

4.2.2
Emitter Spacing and Discharge for a Wetted Strip

In some situations a wetted strip at the soil surface is desired. A wetted strip may be created by means of a line source such as a soaking tube or a drip-tape with virtually continuous outlets. Another possibility is to use a drip line with "discrete" emitters whose discharge and spacing would results in an overlap of saturated radii of water entry ponds. The ultimate saturated radius that develops around a surface emitter is given by (see Eq. 3.13):

$$r_s = \sqrt{\frac{4}{\alpha^2 \pi^2} + \frac{q}{\pi K_s} - \frac{2}{\alpha \pi}} . \tag{4.5}$$

Equation (4.5) relates emitter discharge q and soil hydraulic properties α and K_s with the minimum emitter half-spacing (emitter spacing is $2r_s$) required for overlapping saturated radii. The lateral extent of the saturated strip for the continuous (drip) line source x_s is estimated by the following simple approximation (Warrick 1985):

$$x_s \approx \frac{1}{2}\left[\frac{q_L}{K_s} - \frac{3}{4\alpha}\right], \tag{4.6}$$

where q_L is the line source discharge per unit length $[L^2/T]$. The minimum linear discharge rate for a positive x_s and the onset of a saturated strip is $q > 3K_s/(4\alpha)$. These two approximations were developed for steady-state flow conditions; hence they represent the minimal values for emitter half-spacing inline [Eq. (4.5)], or for minimal half- spacing between line sources for complete surface wetting [Eq.(4.6)]. In other words, for long application times, using smaller emitter spacing and smaller spacing between line sources may result in runoff.

Schwartzman and Zur (1986) proposed geometrical approximations for the lateral extent and depth of the wetted volume under the line source in relation to line discharge and soil saturated hydraulic conductivity. Combining their expressions into a single equation we obtain:

$$d = 1.7z^a\left(\frac{q}{K_s}\right)^b , \tag{4.7a}$$

with the exponent a ranging between 0.75 and 0.85 , and b between −0.15 and −0.25. A reasonable approximation would be:

$$d = 1.7z^{0.8}\left(\frac{q}{K_s}\right)^{-0.2} . \tag{4.7b}$$

This equation may be used to predict the dimensions of the wetted soil developing under a drip line source. In example 4.3 we illustrate the use of Eq. (4.7) for determining minimum spacing between line sources such that adjacent wetted volumes contact each other without overlap.

Example 4.2: Design of Surface Wet Strip Using Discrete Emitters

Problem: Determine emitter spacing for various discharge combinations for a wet surface strip on a soil with $K_s = 0.84\,\mathrm{cm\,h^{-1}}$ and $\alpha = 0.025\,\mathrm{cm^{-1}}$.

Solution: Select several emitter discharge rates and calculate r_s for each q using Eq. (4.5):

Emitter discharge (l/h)	1	2	3	4	8	12
Saturated radius (cm)	6.6	12.0	16.8	21.0	35.2	46.6
Emitter spacing (cm)	13.2	24.0	33.6	42.0	70.4	93.2

Example 4.3a: Minimum Spacing Between Adjacent Line Sources (to attain a completely wet soil surface)

Problem: Determine spacing between line sources to attain a completely wet soil surface for a range of discharge combinations for a soil with Ks = 0.84 cm h⁻¹, $\alpha = 0.025$ cm⁻¹.

Solution: Select a range of line discharge rates and calculate x_s for each q_L using Eq. (4.6):

Line discharge (l/h/m)	1	2	3	4	8	12
Spacing between adjacent line sources (cm)	N/A[†]	N/A	5.7	17.6	65.2	112.8

[†]The minimum line discharge for the onset of a saturated strip for this soil is $q > 3K_s/(4\alpha) = 2.52$ l/h/m.

Example 4.3b: Minimum Spacing Between Adjacent Line Sources for Non-Overlapping Wetted Volumes

Problem: Determine the spacing between line sources for a range of (linear) discharge-root depth combinations in soils with Ks = 1 and 10 cm h⁻¹ (loamy and sandy soils). Maintain non-overlapping wetted soil volumes (in effect you are asked to design for contacting wetted volumes in their largest lateral extent).

Solution: Select a reasonable range of emitter discharge rates and rooting depths and calculate d for the line sources using Eq. (4.7b). Note that the discharge is per unit length (see Table 4.2).

Table 4.2a. Spacing between adjacent line sources (cm) for nonoverlapping wetted volumes in a loamy soil with $K_s = 1$ cm/h

Z_{root} (cm)	Q (l/h/m)				
	1	2	4	8	12
30	40.9	47.0	54.0	62.1	67.3
60	71.3	81.9	94.1	108.0	117.2
90	98.6	113.3	130.1	149.4	162.1
120	124.1	142.6	163.8	188.1	204.0

Table 4.2 b. Spacing between adjacent line sources (cm) for nonoverlapping wetted volumes in a sandy soil with $K_s = 10$ cm/h

Z_{root} (cm)	Q (l/h/m)				
	1	2	4	8	12
30	25.8	29.7	34.1	39.2	42.5
60	45.0	51.7	59.3	68.2	73.9
90	62.2	71.5	82.1	94.3	102.3
120	78.3	89.9	103.3	118.7	128.7

4.3
Hydraulic Design of Laterals

4.3.1
Uniform Slope

After having decided upon the best combination of emitter discharge rate and distance between emitters, we are now able to calculate the length and diameter of the laterals. The relationship between diameter D, length L and flow rate Q, resulting in a certain head loss H_L, is given by the Hazen-Williams formula [Eq. (2.6), Chapter 2]:

$$H_L = 2.78 \cdot 10^{-6} \, FLD^{-4.87} (N\overline{Q}/C)^{1.85} \qquad (4.8)$$

This is an empirical formula, therefore the unit assigned to each dimensioned variable must be retained (see Section 2.2).

In designing the layout of the drip laterals, the lateral length L is normally predetermined by the dimensions of the field. From L and emitter spacing d, as previously determined, the number of emitters N can be computed. Using Eq. (4.8) we can now calculate the tube diameter for a given allowable head loss H_L along the lateral (Bresler 1978). As the spacing between emitters along the lateral is given by d, it is possible therefore to express N = L/d. Substituting L/d for N in Eq. (4.8) and rearranging, Eq. (4.8) is reduced to:

$$L = 88.9 \cdot D^{1.708} (H_L/F)^{0.351} (Cd/\overline{Q})^{0.649} . \qquad (4.9)$$

Equation (4.9) gives the L-D relationships for any prescribed value of head loss H_L for the known Q-d relationship, and assumes F to be constant over any given range of L and d. To select the appropriate value of H_L for Eq. (4.9), a suitable criterion is based on the difference between the emitter discharge at the lateral inlet and the downstream discharge, relative to the average discharge along the lateral, $\overline{Q} \, [L^3 T^{-1}]$:

$$\frac{Q_i - Q_d}{\overline{Q}} \leq \varepsilon, \qquad (4.10)$$

where Q_i is the inlet discharge, Q_d is the downstream discharge and ε is a preselected error fraction, about 0.05.

The relationship between emitter discharge rate and the hydraulic head at the emitter may be given by the empirical expression Eq. (2.5): $Q = kH^x$, with information on the value of the exponent x available from the manufacturer or determined in laboratory tests. Using the maximum value of ε it follows from Eq. (4.10) and Eq. (2.5) that:

$$H_i^x = \frac{\varepsilon \overline{Q}}{k} + H_d^x. \tag{4.11}$$

Since $H_L = H_i - H_d$, then

$$H_i^x = \frac{\varepsilon \overline{Q}}{k} + (H_i - H_d)^x. \tag{4.12}$$

Thus, by knowing the pressure head H at the lateral inlet and the emitter constants k and x, the value of H_L can be calculated from Eq. (4.11) for given x and $\overline{Q}$. This value of H_L is then substituted into Eq. (4.9) to calculate the D-L relationship needed for the lateral design. When the lateral length is given by the size of the plot and H_i is known, the diameter D is calculated from Eq. (4.9). Otherwise, the optimal economic D-L-H_i combination has to be calculated for each field and soil condition.

Another way of tackling this problem is to calculate the length L of a lateral with a given tube diameter D and allowable head loss H_L. H_L must be based on the allowable variation in emitter discharge. A flow variation of 10–15% results in a pressure variation of about 20%, depending on the type of emitter, according to the values of the constants in Eq. (2.5). This criterion of less than 20% pressure variation is called "desirable pressure variation" by Wu and Gitlin (1977). An acceptable pressure variation according to their design criterion is 20–40% resulting in an outflow variation of 15–30%. In addition to the frictional head loss H_L, one also has to take into account the loss or gain in pressure head caused by the direction of the slope H_s, according to: $H_L \pm H_s = \Delta H$ where ΔH is the total pressure drop along the drip lateral. Wu and Gitlin (1977) and Wu et al. (1979) constructed design charts and nomographs which can help the designer select the proper diameter and length of laterals for uniform slopes (see Figs. 4.1 and 4.2).

The procedure for using these figures is as follows:

1. Select design values for input pressure H and length of lateral L and calculate L/H.
2. Find the value of L/H on the vertical axis of quadrant III in Fig. 4.1 and move horizontally in quadrant 1V to the given % slope. Choose the appropriate figure (a or b) depending on whether the slope is upward or downward from the location of the main or submain. From the point of intersection, draw a vertical line into quadrant I at the upper (outside) boundary of the acceptable or desirable region, according to the design criterion. From that point, draw a horizontal line into quadrant II.
3. Draw a vertical line into quadrant II from the calculated L/H value.
4. The intersection of the vertical and horizontal lines in quadrant II will give the value of $\Delta H/L$.
5. Using the nomograph (Fig. 4.2), mark off the value of $\Delta H/L$ (obtained in step 4) and the value of the total discharge per lateral (obtained from the length, the number of emitters and their flow rate Q) and draw a line joining the two points to get the appropriate tube size (the point of intersection of your line and the D axis).

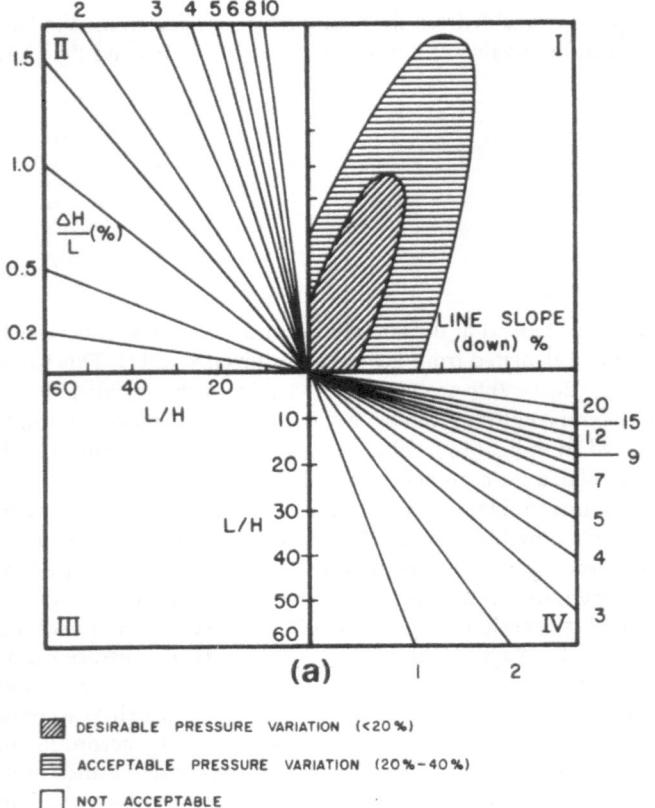

DESIRABLE PRESSURE VARIATION (<20%)

ACCEPTABLE PRESSURE VARIATION (20%-40%)

NOT ACCEPTABLE

Fig. 4.1. General design charts for lateral line and and submain (Wu and Gitlin 1977)

Example 4.4: Use of Design Charts for Laterals and Submains (Figs. 4.1 and 4.2)

Assume a field of length L = 300 m, an inlet pressure of H = 10 m and a down slope of 2%. Taking into account the soil hydraulic properties and the plant water requirements, the best combination of emitter discharge rate and distance between emitters is Q = 3.6 L/h and d = 1 m.

1. These input data give L/H = 300/10 = 30.
2. Draw a horizontal line in quadrant IV of Fig. 4.1a from this point to the intersection of slope 2% down, and from this point draw a vertical line into quadrant I up to the boundary of the desirable and acceptable pressure variation.
3. From this point, draw a horizontal line into quadrant II. This line intersects with the vertical line starting at L/H = 30 at the point of ΔH/L = 2%.
4. The total discharge is (300 m/1 m) 3.6 l/h = 1080 l/h = 0.3 l/s.
5. Draw a line in Fig. 4.2 joining ΔH/L = 2% and Q = 0.3 l/s; it will intersect the middle axis at D = 2 cm.
6. Therefore, the suitable plastic tube diameter is 20 mm.

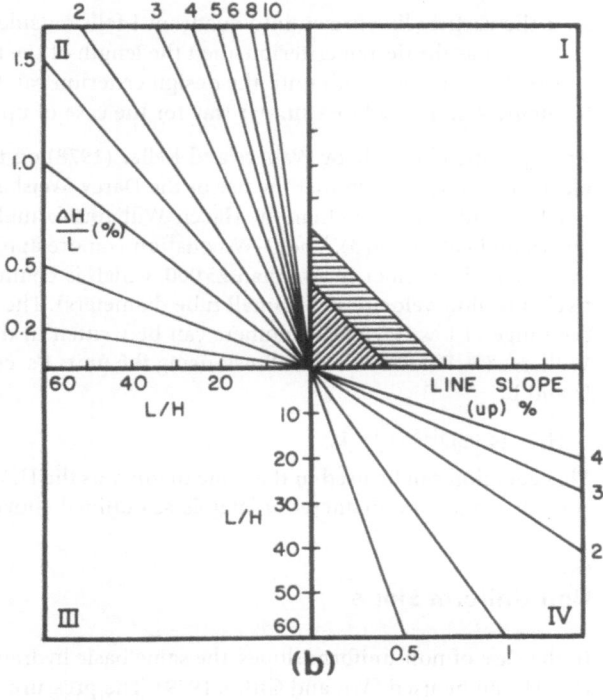

L =   Total length,   meter (m)
H =   Input pressure,   meter (m)
ΔH = Total energy drop by friction at
      the end of line,   meter (m)
$\frac{\Delta H}{L}$ = Total friction drop and
      length ratio, %

Fig. 4.1. *Continued*

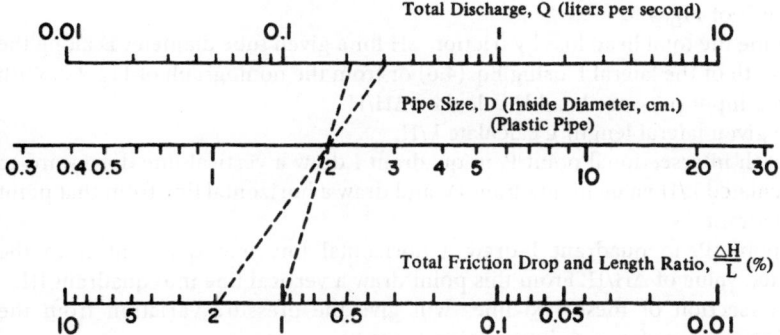

Fig. 4.2. Nomograph for drip irrigation laterals and submain design (Wu et al. 1979)

7. If the vertical line drawn into quadrant I falls outside the zone of pressure variation set as the design criterion, then the length of the field L has to be divided into smaller units of length until the design criterion can be met.
8. Figure 4.2b is used in a similar way for the case of upward slopes.

An experimental study by Watters and Keller (1978) on friction head loss in smooth plastic tubes has shown that the use of the Darcy-Weisbach (D-W) equation resulted in a better fit to the data than the Hazen-Williams formula as was originally assumed by Wu and Gitlin (1973). The D-W equation is more appropriate in the region where the Reynolds number is less than 25 000, which is common in drip irrigation (relatively low flow velocities and small tube diameters). The D-W equation calculated in the range of low Reynolds numbers can be written in the same form as the Hazen-Williams (H-W) formula as follows, using the units l/s, cm and m, respectively for Q, D and L,

$$H_L = 14.03 Q^{1.85} D^{-4.87} L. \tag{4.13}$$

This equation can be used in the same manner as the H-W equation in order to select suitable length and diameter of laterals as outlined above.

4.3.2
Non-Uniform Slope

In the case of non-uniform slopes, the same basic hydraulic concept as given by $H_L \pm H_s = H$, can be used (Wu and Gitlin, 1979). The pressure variation along the lateral is determined by a linear combination of head loss by friction and head loss or gain by change in elevation (slope). The lateral can be divided into sections of relatively uniform slope, and ΔH can be calculated separately within each section. A dimensionless design chart was developed along these principles by Wu and Gitlin (1979) for use with nonuniform slopes (Fig. 4.3). The procedure for the use of this chart is as follows:

1. Divide the nonuniform slope length L into several sections of length l_i, each of which can be considered to have uniform slope. Determine the head loss or gain by slope in each section, and calculate the cumulative head gain or loss for all the points along the lateral (ΔH_i).
2. Plot the nonuniform slope pattern in a dimensionless form, i.e. $1/L$ vs. $\Delta H_i/L$ in quadrant I of Fig. 4.3.
3. Determine the total head loss by friction ΔH for a given tube diameter D along the total length of the lateral L using Eq. (4.6) or from the nomograph of Fig. 4.2. With the given input pressure head H calculate $\Delta H/H$.
4. For the given lateral length L calculate L/H.
5. From each intersectional point P_i in quadrant I draw a vertical line downward to the calculated L/H value in quadrant IV and draw a horizontal line from that point into quadrant III.
6. From point P_i in quadrant I draw a horizontal line into quadrant II to the calculated value of $\Delta H/H$. From this point draw a vertical line into quadrant III.
7. The intersection of these two lines will give the pressure variation from the operating pressure for each intersection point Pi.

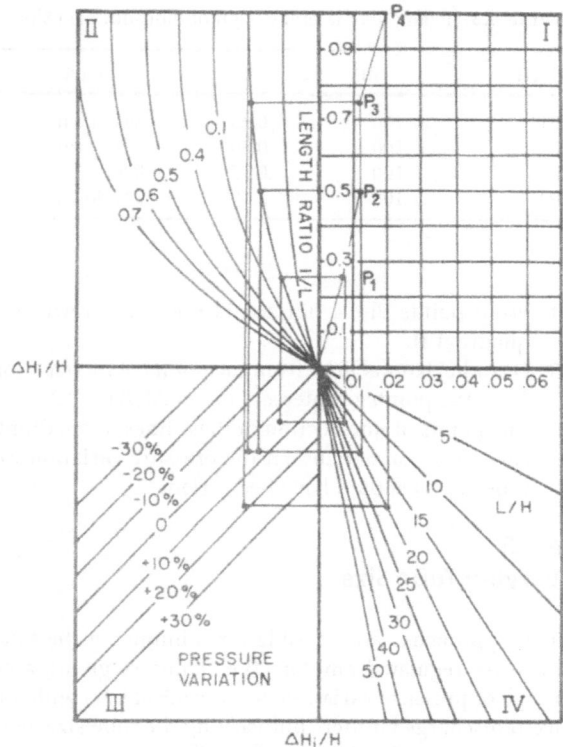

Fig. 4.3. Design charts for non-uniform slopes (Wu and Gitlin 1979)

8. If the pressure variation is within the design criterion (for instance ±10% as a desirable design), then one can try a smaller tube diameter D and see if it still meets the design criterion. If, on the other hand, the obtained pressure variation is outside the limit of the design criterion, then a larger tube diameter D must be considered.

Example 4.5: Using Design Charts for Laterals on Non-uniform Slopes (Fig. 4.3)
Assume a field of length L = 400 m, initial H = 20 m and Q = 0.22 l/s (emitters of discharge 2 l/h spaced one meter apart). The length of the field can be divided into four equal sections (of length 100 m) with the following slopes: 3% down, 2% down, no slope and 3% down.

1. The following table can be constructed based on these data (Table 4.3).
2. Points P1 to P4 are plotted in quadrant I. Assume a tube with a diameter D of 20 mm, then ΔH is determined using Fig. 4.2; take Q = 0.22 l/s. Then $\Delta H/L = 0.01$, $\Delta H = 4$ m and L/H = 400/20 = 20.
3. From points P1–P4 vertical lines are drawn to intersect line L/H = 20 in quadrant IV. $\Delta H/H = 4/20 = 0.2$.

Table 4.3. Example of lateral design on non-uniform slope

Section	1 (m)	1/L	Slope (%)	H_I	ΔH_i	$\Delta H_i/L$
P1	100	0.25	3% down	3	3	0.0075
P2	100	0.50	2% down	2	5	0.0125
P3	100	0.75	0 %	0	5	0.0125
P4	100	1.00	3% down	3	8	0.0200

4. From points Pl–P4 horizontal lines are drawn to intersect curve $\Delta H/H = 0.2$ in quadrant II.
5. Draw horizontal lines from the points of intersection on $L/H = 20$, and vertical lines from the points of intersection on $\Delta H/H = 0.2$.
6. The points of intersection of these lines in quadrant III are in the permissible pressure variation of ±20%. If the pressure variation would have been greater, a larger tube size D should have been tried.

4.3.3
Varying Tube Size

Two approaches are possible for minimizing the total cost of the system: (1) using pressure regulated emitters (i.e., emitters giving a constant discharge over a large range of pressure variation, usually 0.5–4 bar) with a fixed lateral diameter; (2) using fixed discharge emitters but varying the tube size along the lateral to minimize pressure variation. The first approach is recommended for use on difficult terrain (steep, varying slopes) enabling the use of small diameter laterals. With the second approach, the pressure variation resulting from slope and frictional head loss along the lateral can be regulated.

It has been shown by Wu and Gitlin (1979) that the energy gradient line for a lateral with a given diameter may be described by an exponential-type curve. If the frictional head loss is balanced by the head gain through elevation change, the maximum pressure variation will be 0.36 SL, S being the slope and L the length of the lateral. This maximum will occur near the middle of the lateral section. When a series of lateral diameters is used, this pressure variation can be reduced considerably. For instance, if four equal sections are used with different diameters, the maximum pressure variation will be only 0.09 SL (Wu and Gitlin 1979). This approach can be used for uniform or non-uniform downhill slopes, causing a modification of the Hazen-Williams formula:

$$S = 15.27Q_m^{1.85} D^{-4.87}, \tag{4.14}$$

where S is the slope of a section in m/m (note that the head gain by slope is balanced by the frictional head loss), Q_m is the mean discharge of the given lateral section in l/s and D is the inside diameter of the lateral section in cm. Nomographs for solving Eq. (4.14) (Wu and Gitlin 1977) are given in Fig. 4.4. The procedure for a variable lateral diameter design in a downslope situation is very simple. The length of the lateral is divided into several sections of equal slope S. The mean discharge for each section Q_m is calculated, beginning at the last section. The

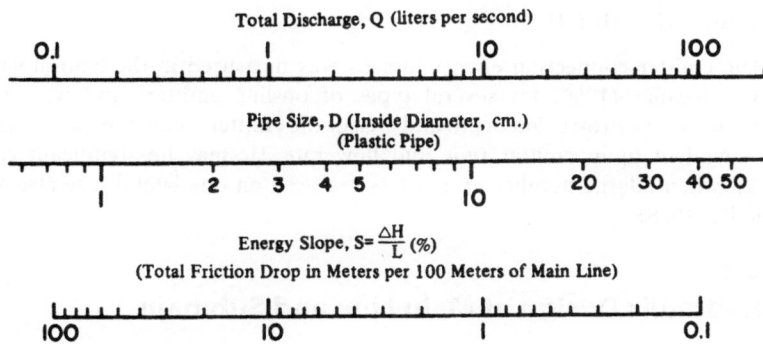

Fig. 4.4. Nomograph for drip irrigation main-line design (Wu and Gitlin 1979)

Table 4.4. Example of lateral design with varying tube diameter

Section	Q (l/s)	S (%)	D (cm)
1	0.250	5	20
2	0.225	5	20
3	0.200	3	20
4	0.175	3	20
5	0.150	1	20
6	0.125	1	20
7	0.100	3	16
8	0.075	3	12
9	0.050	5	8
10	0.025	5	8

required tube diameter D can now be read for each section from the nomograph of Fig. 4.4.

Example 4.6: Design for a Variable Tube Diameter
Take, for example, an apple orchard with 4 m spacing between trees and each row 500 m in length. Each tree is irrigated with two emitters of Q = 3.6 l/h. The downslope varies in a discrete manner approximately every 100 meters as follows: 5, 3, 1, 3 and 5%, respectively. The length of the row is divided into ten sections of 50 m. The discharge Q for each section is calculated starting at the bottom section (as given in Table 4.4). The required tube diameter D can be read from Fig. 4.4 by matching Q with S.

4.3.4
Energy Loss Across Emitter Connections

Howell and Barinas (1980) have shown that, apart from the frictional head loss H_L [Eq. 4.13)] along the lateral an additional head loss H_E at the connections of on-line emitters to the lateral must be taken into account. Therefore, the total pressure drop ΔH is given by

$$\Delta H = H_L + H_E \pm H_S. \tag{4.15}$$

The emitter connection energy loss H_E was measured in the laboratory by Howell and Barinas (1980) for several types of on-line emitters and was shown to be related to the protrusion depth and area of the emitter connections. H_E was also highly dependent upon emitter type and flow rate. H_E may be significant compared to ΔH when a large number of emitters are used on one lateral (see also Watters and Keller 1978).

4.4
Hydraulic Design of Main Line and Submain

4.4.1
Design of Main Lines and Submains

The hydraulic design of main lines and submains for drip irrigation is based on the same principles as for laterals, i.e., topography, the available pressure at the head of the system, and the required pressure and discharge at each outlet (submain and lateral). The procedure is as follows (see Fig. 4.5 for two examples):

1. Plot the main-line profile, divided into sections according to outlets and average slope for each section.
2. Add to the obtained profile the pressure needed for irrigation at the head of each lateral.
3. Mark the available input pressure and draw a straight line from this point to the end of the line. If the input pressure is not high enough, then draw several sections of straight lines.
4. Tabulate the required discharge Q_m and the energy slope S of each section. With the help of Eq. (4.14) or the nomograph of Fig. 4.4, calculate the pipe diameter for each section.
5. Wu (1975) demonstated that this simple straight energy line method leads to designs similar to the optimal obtained by computer simulation.

4.4.2
Computer Aided Hydraulic Design

Many drip irrigation companies and dealers employ computer software packages to aid with the hydraulic design of the irrigation system, including pipe sizing, length, valves etc., and to develop hydraulic tables for the various options. These powerful design tools are capable of compiling lists of equipment quantities and costs and usually provide a graphical output of the system layout. Some of the advanced packages incorporate optimization routines for minimizing costs, or head losses, while solving for the hydraulic performance of the entire pipe network. The user may select from databases on the hydraulic characteristics of different products (as supplied by various manufacturers). These databases contain descriptive, dimensional, cost, and technical information about the components used in each design. Two examples of widely used design programs are: IRRICAD (AEI Software, New Zealand), and WCADI (Weizman Industries, LTD, Israel). It is always good practice to use the hydraulic

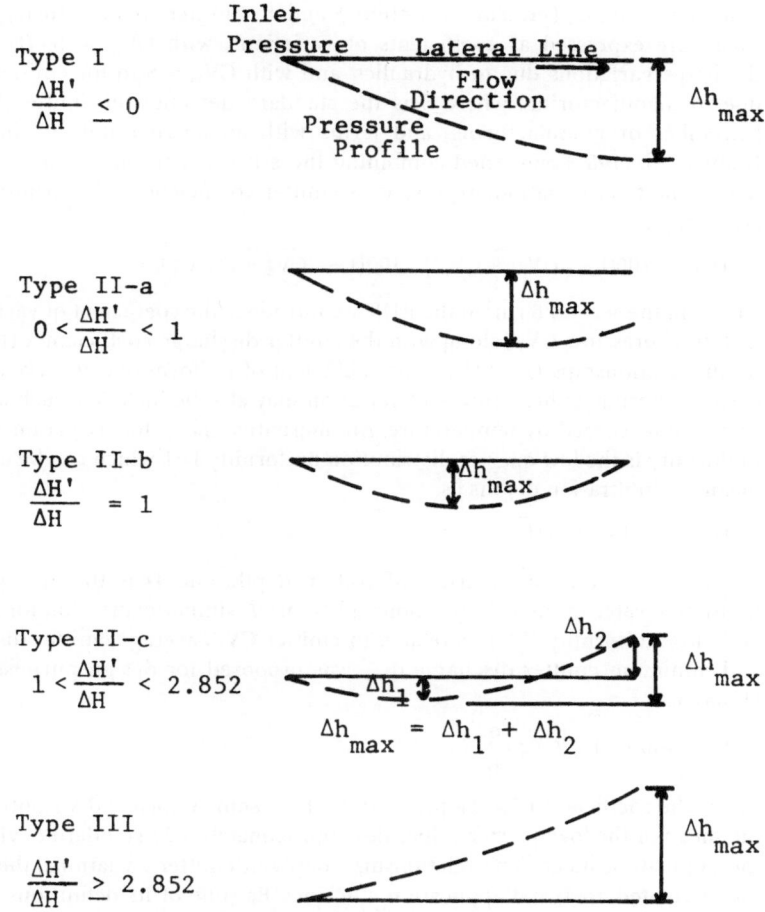

Fig. 4.5. Energy gradient lines for different field situations (Wu 1975)

design principles discussed above for spot-checking design recommendations obtained by these complex programs.

4.5
Effects of Variability in Hydraulics, Soil and Plant Properties

4.5.1
Effects of Emitter and Hydraulic Variability

The objective of a well-designed drip irrigation system is to supply the crop with uniform amounts of water throughout the field. Variability in emitter discharge affects the total amount of water applied at each location, resulting in variability in plant-available soil water. Emitter variations may be attributed to variations due to system hydraulics (pressure variations) and to emitter manufacturing variations. These are expressed as coefficients of variations, with $CV_H{}^x = S_H{}^x/\bar{H}{}^x$ for emitter discharge variations due to hydraulics, and with $CV_m = S_q/\bar{q}$ for emitter variations due to manufacturing, where S is the standard deviation of emitter flow due to hydraulics or manufacturing, and terms with an overbar indicate mean values. Bralts et al. (1981) suggested combining these two coefficients of flow variation to obtain the total variation, expressed as emitter coefficient of uniformity UC_{em} [%], according to

$$UC_{em} = 100\left(1 - \sqrt{CV_{H^x}^2 + CV_m^2}\right) = 100\left(1 - \sqrt{CV_m^2 + x^2 CV_H^2}\right), \tag{4.16}$$

where, in the second term on the RHS, we introduce the coefficient of variation of the hydraulic pressure, CV_H, along with the emitter discharge coefficient x (this is based on the relationships $Q = kH^x$). This coefficient of uniformity can serve as a general design criterion. Other sources of variation may also be included, such as the variation in flow caused by temperature. An alternative index for irrigation distribution uniformity is the low quarter distribution uniformity DU, which can be expressed by means of infiltration depths as:

$$DU = 100\left(1 - LQD/\bar{D}\right), \tag{4.17}$$

where LQD is the low quarter infiltration depth, and $\bar{D}$ is the average depth of infiltrated water in the field (or along a lateral). A similar expression for low quarter emission uniformity (EU) as related to emitter CV_m, average emitter discharge q_{avg}, and minimum emitter discharge q_{min}, was proposed for design purposes by Bralts (1983) as:

$$EU = 100(1 - 1.27 CV_m)\frac{q_{min}}{q_{avg}}, \tag{4.18}$$

where the coefficient 1.27 is a property of the assumed normal distribution in which the mean of the low quarter values lies approximately 1.27 standard deviations from the population mean. The relationships between emitter variations, the portion of underirrgated area, and application efficiency Ea (one of its definitions is: q_{min}/q_{avg}) are summarized in Fig. 4.6 from Wu et al. (1986).

Equations (4.17) and (4.18) suggest a link between emitter discharge uniformity and resultant variations in depths of infiltrated water. A uniformity coefficient for the infiltrated depths (or amounts of water infiltrating at each location) is:

$$UC = 100\left(1 - |D|/\bar{D}\right), \tag{4.19}$$

where $|D|$ is the average absolute deviation from mean infiltrated depth $\bar{D}$. This coefficient is identical to the uniformity coefficient of Christiansen (1942), developed for sprinkler irrigation. Warrick (1983) established relationships between UC and DU and the statistical coefficient of variation CV (= standard deviation/mean) for various distributions. These relationships for various distributions are presented in Fig. (4.7) and are approximated as follows:

$$UC = 100(1 - 0.8 CV), \tag{4.20}$$

$$DU = 100(1 - 1.3 CV), \tag{4.21}$$

$$DU = -60 + 1.6 UC. \tag{4.22}$$

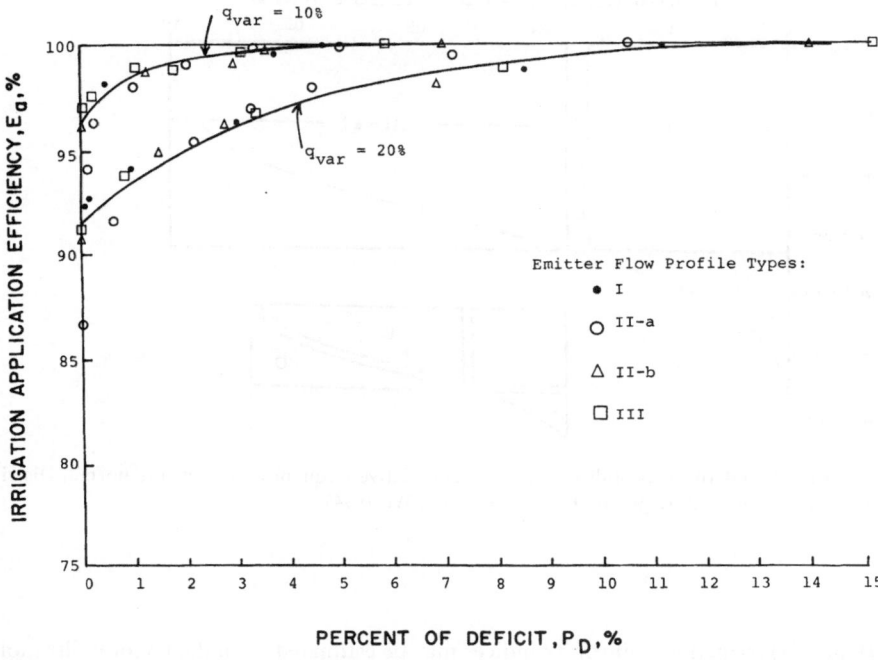

Fig. 4.6. The relationship between irrigation application efficiency and deficit for emitter flow variation by hydraulics, q_{var} of 10 or 20% (Wu et al. 1986)

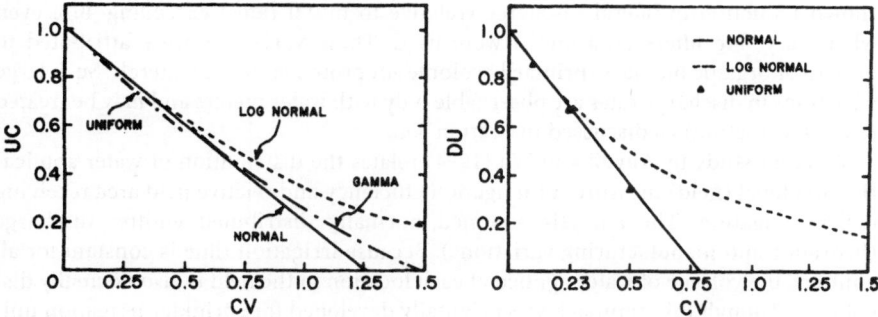

Fig. 4.7. Relationship between the uniformity coefficient (UC) and the coefficient of variation (CV) for several distributions and the distribution uniformity of the low-quarter (DU) and the coefficient of variation (CV) (After Warrick 1983)

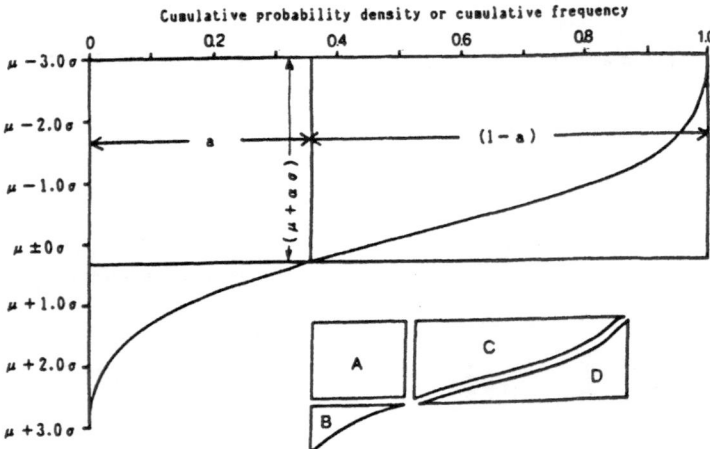

Fig. 4.8. Cumulative probability density or cumulative frequency curve of the normal Distribution. μ = required irrigation depth (Anoyi and Wu 1994)

Hence, (1) irrigation uniformity indices may be estimated from the CV of infiltration depths; and (2) the detailed shape of the distribution is not important for CVs of less than 0.25. Wu (1995) points out that under most conditions manufacturer variations and hydraulic design are less significant factors affecting irrigation uniformity (as long as designed within a specified range), and emitter clogging is likely to be the most significant factor. This assessment is corroborated by the findings of Hanson et al. (1995) in a recent survey and field system tests. Studies by Ravina et al. (1992) have shown reduction in lateral discharge (relative to initial rates) exceeding 40% even when very fine filters (120 mesh) were used. These variations were attributed to growth of organic biomass (primarily colonies of protozoa) within laterals. Such large variations in discharge rates are observable only with water meters and may be treated (e.g., chlorination) as discussed in Section 4.4.

A recent study by Anyoji and Wu (1994) relates the distribution of water application to global (field) measures of irrigation efficiency and relative field area receiving deficit irrigation. The analysis assumed normally distributed emitter discharge (hydraulic and manufacturing variations). Because irrigation time is constant for all emitters, the volume of water applied at each location in the field is also normally distributed. Though this approach was originally developed for sprinkler irrigation uniformity measures (i.e., a one-dimensional flow regime) by Hart and Reynolds (1965) and Walker (1979), it has some merit in the context of drip irrigation, at least at the field scale. The key to the approach is using known statistical moments of q (i.e., mean μ_q and standard deviation σ_q), and a prescribed required irrigation depth $y = \mu_q + \alpha\sigma_q$, to derive explicit expressions for the different areas under the cumulative probability density function (see Fig. 4.8 which is also Fig. 2 of Anyoji and Wu 1994). These areas represent the amount stored in the root zone (A + C), the deep percolation (B), and the amount of deficit (D), and are directly related to various efficiency measures.

At a field scale, these areas are interpreted as fractions of emitters delivering the respective amounts of irrigation water (or the amounts received at various locations in the field).

Considering the transformation $u = (y - \mu_q)/\sigma_q$, which standardizes the normal distribution to mean zero and a standard deviation of 1 [$N(0,1)$], the area under the cumulative distribution is

$$a(\alpha) = \int_{\alpha}^{\infty} \frac{1}{\sqrt{2\pi}} \exp^{-\frac{u^2}{2}} du. \tag{4.23}$$

The value of $a(\alpha)$ may be estimated from tables or provided by closed-form approximations[1]. The areas in Fig. 4.8 are computed, and an expression for irrigation application efficiency, Ea (defined as the percentage of the total amount of irrigation applied that is stored in the root zone available for crop use):

$$Ea = \frac{A+C}{A+B+C} \times 100 = \left(1 + a(\alpha)\alpha CV_q - \frac{1}{\sqrt{2\pi}} \exp^{-\frac{\alpha^2}{2}} CV_q\right) \times 100, \tag{4.24}$$

where $CV_q = \sigma_q/\mu_q$ (which may be expressed using emitter coefficient of uniformity given in Eq. 4.16 as: $CV_q = 1 - [UC_{em}/100]$). Similarly, the percent deficit, PD, defined as the ratio of the amount of deficit to the total amount required (Anyoji and Wu 1994) is given by:

$$PD = \frac{D}{A+C+D} \times 100 = \left(\frac{[1-a(\alpha)]\alpha CV_q + \frac{1}{\sqrt{2\pi}} \exp^{-\frac{\alpha^2}{2}} CV_q}{1 + \alpha CV_q}\right) \times 100. \tag{4.25}$$

The relationships between PD and Ea for different values of emitter variability are given in Fig. 4.8 (Note the symbols were obtained for different lateral pressure profiles; see Fig. 4.5).

These expressions were used by Anyoji and Wu (1994) to devise an irrigation schedule for the special case where $\alpha = 0$ (i.e., the amount applied equals the amount required and no additional amount of water is scheduled to compensate for system nonuniformity). The resulting application efficiency (Ea) was always over 92% and PD was less than 8% whenever $CV_q < 0.2$ (Ea = [$1 - 0.4\ CV_q$] $\times$ 100).

4.5.2
Effects of Soil Spatial Variability and Plant Root Zone

The volume of water discharged by an emitter at a location is affected by the emitter hydraulic characteristics that may vary along a lateral or between laterals. These variations result in variations in wetted soil volumes even when the soil properties are uniform. In many situations, however, variations in soil hydraulic properties are superimposed on variations in emitter discharge rates, leading to nonuniform infiltration, water retention, and water availability to the crop. A schematic illustration

---

[1] See Abramowitz and Stegun (1964, p. 932) for their expression (26.2.18) where $a(\alpha)$ is approximated as: $a(\alpha) = 0.5/(1 + c_1\alpha + c_2\alpha^2 + c_3\alpha^3 + c_4\alpha^4)^4$ with: $c_1 = 0.196854$; $c_2 = 0.115194$; $c_3 = 0.000344$; $c_4 = 0.019527$

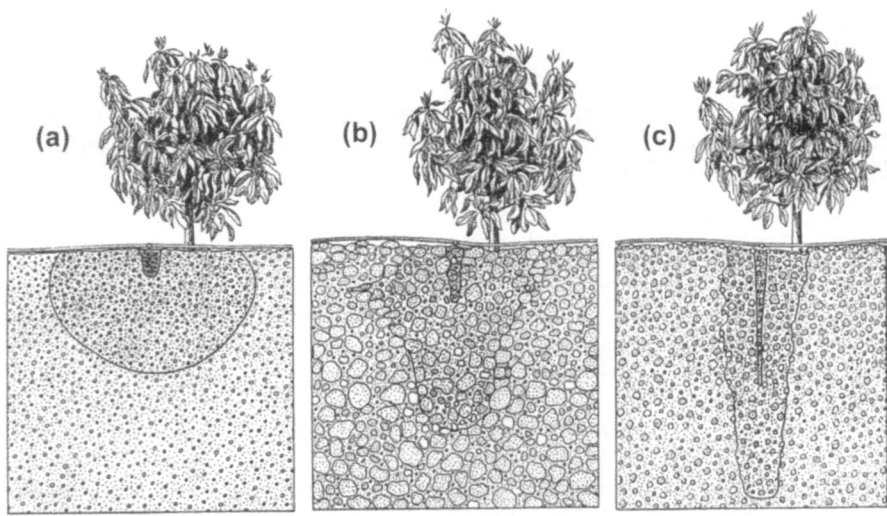

Fig. 4.9. Wetting patterns for drip irrigation of soils with different textural and hydraulic properties (Shoji 1977)

showing the effect of different soil properties on wetting patterns under drip irrigation is depicted in Fig. 4.9 (Shoji 1977).

Whenever possible, partitioning the field into relatively uniform management units (in terms of their soil properties) should result in a better match between drip system design and soil properties. In many situations, however, soil properties vary from location to location erratically with no apparent trend. Attempts have been made to incorporate spatial statistical information on soil nonuniformity into drip irrigation design using a geostatistical framework (Russo 1983, 1984). The resulting design is a complex map of emitter discharge rates and spacing for different locations in the field matching the pattern of soil variations. Though theoretically feasible, such a complex design requires a substantial amount of detailed localized soil information (i.e., knowledge of soil properties at many locations in the field) which is costly and very difficult to implement on a large scale due to technical, agronomic, and management difficulties.

In principle, concepts similar to those discussed by Anyoji and Wu (1994) may be applicable to quantifying effects of soil variability by assuming a certain distribution of soil intake or storage properties. If emitter discharge may be considered uniform, spatial variations in soil properties result in differences in the resultant field-wide distribution of the amounts of plant available soil water. There is, however, an important aspect of drip irrigation that could complicate such analysis, and this is the partial and three-dimensional wetting patterns. Such partial wetting becomes particularly important for field crops where the combination of emitter spacing, discharge rates, and soil properties affect the availability and accessibility of soil water by plant roots. These aspects of plant root zone-scale non-uniformity have received relatively little attention. Only a few studies have attempted to address the complex relationships between emitter spacing and variations in wetting patterns and the horizontal extent of crop rooting zones (Seginer 1979; Cogels 1983; Wallach 1990); however, none of these studies incorporate effects of soil variability. In assessing the relationships between (sprinkler) irrigation uniformity and plant root systems, Seginer (1979) concluded that drip irrigation is likely to result in high *effective* uniformity, whereas the detailed spatial distribution is very non-uniform. The term *effective* uniformity represents the distribution of soil water among the rooting zones of individual plants. An example of high and low effective uniformities for furrow vs. drip-irrigated sugar cane is illustrated in Fig. 4.10 (Bui and Kinoshita 1985). It is interesting to note that even when a uniform *effective* distribution is attainable from hydraulics and layout considerations, soil variability can significantly alter the actual *effective* uniformity.

Incorporation of soil variability considerations into drip irrigation design clearly requires a certain amount of soil sampling and characterization. For example, a relatively simple and informative soil attribute is soil texture that contains information on soil hydraulic properties, and potentially, their variability. When textural (or similar) information is available, one might identify a few classes of soil properties/attributes (e.g., "sandy", "clayey", "stony"), and consider the relative area represented by each class within the irrigated field (note, these are not continuous parcels of land). Assuming that each soil "class" requires different emitter discharge and spacing design, the question of finding an optimal design for the entire field is reduced to an optimization process using the relative areas of each class as weight functions for each alternative design. A similar optimization methodology has been implemented for optimal irrigation management in heterogeneous fields (Or and Hanks 1993). The implementation of such methodologies requires better understanding of the relationships between the effects of root-zone non-uniformity (induced by soil type, emitter discharge and extent of root zone) and the resulting crop yield.

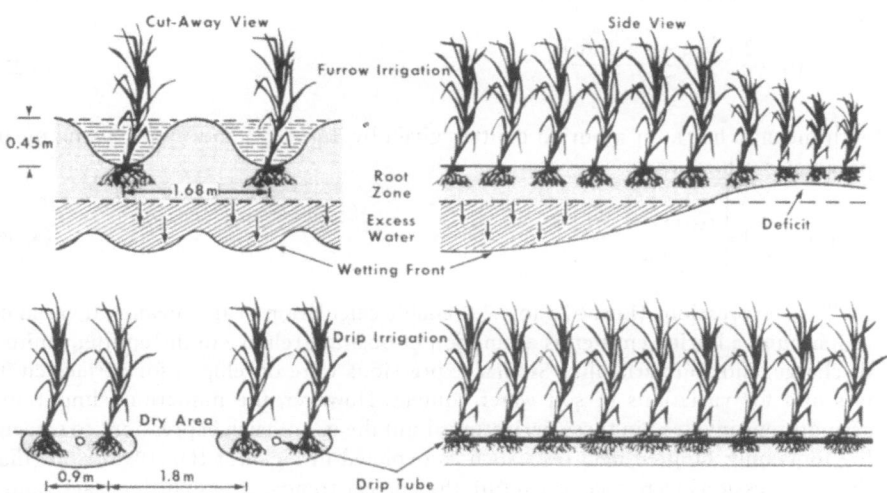

Fig. 4.10. Furrow irrigation vs. drip irrigation of sugarcane in Hawaii. Water distribution uniformity with furrow irrigation is 30%, with drip 80% (Bui and Kinoshita, 1985)

4.5.3
Influence of Soil Spatial Variability on Monitoring Soil Water Status

Large variations in soil properties (even for uniform emitter discharge, such as that obtainable with pressure compensating emitters) affect the ability to reliably monitor soil water status by using sensors buried in the soil. Apart from the selection of a proper location relative to the dripper and the crop row for sensor placement, large variations in water content and matric potential may exceed the range of operation for certain sensors (e.g., tensiometers). Several studies were conducted to understand the extent and patterns of spatial and temporal variations in water content and matric potential within drip irrigated fields (Hendrickx and Wierenga 1990; Or 1995a, 1996). Unlike soil water monitoring with most other irrigation methods, non-uniform distribution from an emitter requires careful consideration of sampling distance relative to the emitter. Two studies aimed at relating spatial variations in soil hydraulic properties (K_s and α) to soil water content and matric potential were conducted by Or (1995a, 1996). Steady state analytical solutions for water flow from point sources were used as the basis for analysis along with statistical representation of variations in soil hydraulic properties (in terms of their means, variances and spatial covariances).

An example from Or's (1995) analysis relates known mean values of $Ks = \bar{K}_s$ and $\alpha = \bar{\alpha}$, and their variances (σ^2_{Ks} and σ^2_{α}), to the mean and variance of the resulting matric potential near the emitter. The expressions are derived for known emitter discharge (q), steady state flow conditions, and for relatively mild soil variation (i.e. small σ^2_{Ks} and σ^2_{α}). The variance of the matric potential (h) around a buried emitter is given as a weighted sum of the variances of the two soil properties (ignoring correlation between K_s and α) as:

$$\sigma^2_h(\rho) = A^2_\alpha(\rho)\sigma^2_\alpha + A^2_Y\sigma^2_Y, \tag{4.26}$$

where $Y = \ln(K_s)$, $\rho = (r^2 + z^2)^{1/2}$. The weight functions are: $A_Y = 1/\bar{\alpha}$, and $A_\alpha(\rho)$, given by:

$$A_\alpha(\rho) = \frac{2 + \bar{\alpha}[z - \rho - 2\bar{h}(\rho)]}{2\bar{\alpha}^2}, \tag{4.27}$$

with mean h, $\bar{h}(\rho)$, for a buried emitter, given by Eqs. (3.8) and (3.12) (using mean parameter values) as:

$$\bar{h}(\rho) = \frac{1}{\bar{\alpha}}\ln\left[\frac{\bar{\alpha}q\exp^{\frac{\bar{\alpha}}{2}(z-\rho)}}{4\pi K_s\rho}\right]. \tag{4.28}$$

These expressions (Eqs. 4.26 to 4.28) enable calculation of the mean and variance of h around a buried emitter as a function of position relative to the emitter, emitter discharge, and soil variability. Similar expressions were developed for surface emitters and for variations in soil water content. However, the numerous simplifying assumptions involved in these derivations limit the use of such expressions to screening tools only. Limited field tests such as depicted in Fig. 4.11 (Or, 1996) show that these expressions were able to capture the correct trends, although the exact values may be different when emitter discharge variability and plant root uptake are added to the picture.

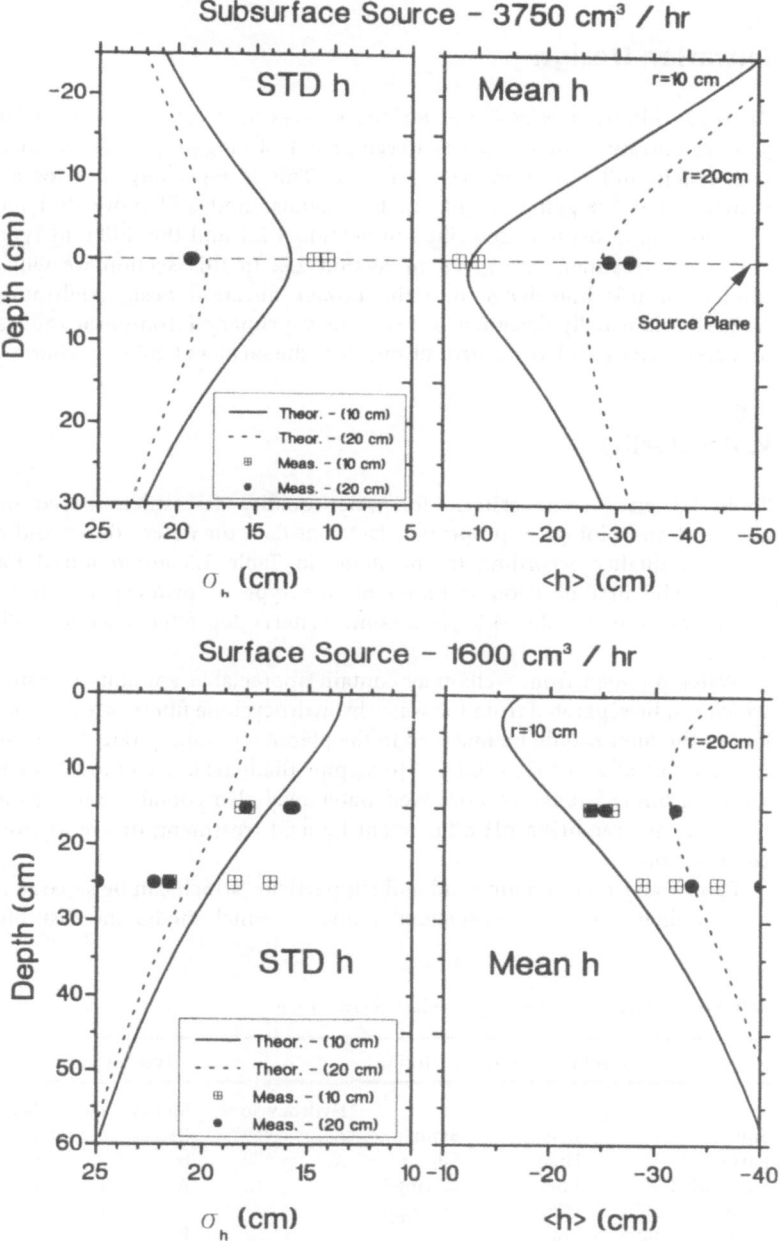

Fig. 4.11. Comparison between model predictions and measurements of steady-state matric head means and standard deviations for subsurface and surface emitters (Or 1996)

4.6
Filtration Design

Adequate filtration is essential for the successful design and operation of a drip system. Filtration failure means development of clogging which, in many cases, is irreversible and results in system failure. This is especially true for a subsurface system, where clogging is difficult to diagnose and still more difficult to repair. The clogging process is described in Section 1.2.1 and the different types of filters and water treatment are given in section 2.4. In this section we will attempt to implement this knowledge into the proper filtration design. Filtration for drip irrigation is usually designed in two stages: primary filtration at the water source and secondary check or control filtration at the farmers field or orchard.

4.6.1
Water Quality

Table 1.3 gives some criteria for water quality assessment based on physical, chemical and biological properties. Detailed data on water source and an analysis of water quality according to the items in Table 1.3 are required for filtration design. The first decision is based on the type of primary filtration, based on the water source. Table 4.5 gives some criteria for filter selection (Plastro Gvat 1989).

    Water pumped from wells may contain appreciable amounts of sand or gravel, which can be separated from the water by hydrocyclone filters (see Section 2.4.1). The size of the filter should be matched to the planned pumping rate. Filters are available in the range of 2 to $300 \, m^3/h$ (3/4″ to 8″ pipe diameter). Each model is effective only within a limited range of flow. Well water may also contain excess concentrations of carbonate, requiring pH adjustment by acid treatment; or excess iron requiring chlorination.

    River water may contain sand and silt particles, which can be separated by hydrocyclone filters, as well as suspended solids, for which media and disk filters are the

Table 4.5. Selection of filter type (Plastro Gvat 1989)

Factor	Contamination	Criterion	Type of filter				
			Hydrocyclone	Media	Disc	Screen	Check
Soil	Low	Sand	A	B	–	C	Screen
Particles	High	Silt	A	B	–	C	Screen
Suspended	Low	<50 mg/l	–	A	B	C	Disc
solids	High	>50 mg/l	–	A	B	–	Disc
Algae, organic	Low		–	B	A	C	Disc
material	High		–	B	A	C	Disc
Iron and	Low	<0.5 mg/l	–	B	A	A	Disc
Manganese	High	>0.5 mg/l	–	A	B	B	Disc

A is the recommended alternative
B is the second choice
C is the third choice

best solution (see Section 2.4.2). Gravel or media filter tanks are available for flow rates of approximately 5–50 m³/h (30–120 cm tank diameter). For higher flow rates a battery of several tanks in parallel should be installed.

Water from reservoirs is the most problematic for drip irrigation, as it contains not only suspended solids, but also appreciable amounts of algae, zooplankton, bacteria and other organic material. Filtration by media or disk filters is insufficient, and water treatment by chloride is essential (Ravina et al. 1992, Tajirshi et al. 1994).

If water of drinking quality is supplied to the grower, the filtration becomes simpler and disk or screen filters may be sufficient in this case. They are available in a wide range (3 to 50 m³/h) and for different filtration grades (20–800 μ). For most drip systems 200 μ (≈80 mesh) is sufficient, but many designers prefer 100 μ (≈150 mesh). The suppliers of many screen filters give data on the filter area. As a rule of thumb, 60–100 cm² is necessary for 1 m³/h of flow, roughly equivalent to a filter volume of 100 cm³ for disk filters.

Secondary or check filtration should be carried out at the farmer's field, generally with screen or disk filters. Disk filters are preferable since they are less prone to mechanical damage and the cleaning by manual or automatic backflushing is more complete, particularly in the case of water from reservoirs or slow flowing rivers. The larger size algae and other organisms are filtered out by the primary filter, but the small particles could develop in the main line into bacterial slime and may cause serious clogging problems.

4.6.2
Water Treatment

The principles of chemical water treatment were given in Section 2.4.5. Here we will quantify this information for design purposes. The need for chemical water treatment is determined by the water quality.

Acid treatment is necessary for the prevention of calcium and/or magnesium carbonate precipitation, which may occur at high pH and at increased temperatures. The pH should be adjusted to a level of 6.0. For most waters with high pH, 1 me/l of acid is sufficient (Nakayama 1986). With continuous dilute acid treatment, carbonate precipitation can be prevented. Iron and manganese sulfide precipitations need higher acid concentrations.

Chlorination is necessary to prevent the growth of bacterial slime and other microorganisms. Adequate chlorination is achieved when the residual concentration of chlorine at the end of the laterals is not less than 1 ppm. This can be achieved by the addition of 1 kg of chlorine gas to 1 m³ of water, or 10 l of sodium hypochlorite with 10% available chlorine to 1 m³ of irrigation water (Nakayama 1986).

The chemical injectors to be used, some of which are described in Section 2.5, should be resistant to corrosion which may be caused by the chemicals. The rate of injection of the chemical q_c (in l/h) should be adjusted to the system flow rate according to

$$q_c = K(dQ_s/c),$$ (4.29)

where K = 3.6×10^{-3} (a conversion constant) d = desired dosage in irrigation water (ppm), Q_s = irrigation system flow rate (l/s), and c = the concentration of the component in the liquid chemical concentrate in kg/l (Keller and Bliesner 1990). The size of the tank containing the concentrated chemical should be adjusted to the area irrigated by the system.

4.7
Fertigation Design

Fertigation is a prerequisite for drip irrigation. Since the wetted soil volume is limited, the root system is confined and concentrated. The nutrients from the root zone are depleted quickly and a continuous application of nutrients along with the irrigation water is necessary for adequate plant growth. Fertigation offers precise control on fertilizer application and can be adjusted to the rate of plant nutrient uptake.

The first step in fertigation design is the choice of the fertilizing system (several are described in Section 2.5 and in Fig. 2.12). This choice depends on the availability of electricity at the irrigation head, the water pressure and the presence of an irrigation controller. The simplest fertilizing system is a fertilizer tank, with no need for external power, only a small pressure loss and easy operation. The main disadvantage is that it must be filled with fertilizer before each water application. If no electrical power is available, than the Venturi or the hydraulic injection pump can be used, provided the water pressure is high enough. These devices need a minimum pressure for operation and the pressure loss, especially for the Venturi, can be considerable. If electrical power is available, than an electrical injection pump is the best solution. Piston pumps are less variable with pressure than diaphragm pumps. All these devices, except the fertilizer tank, can be equipped with timers or flow meters that can be attached to an irrigation controller. They can be set to operate at a certain rate of injection and during part of the irrigation cycle, providing an adequate volume of water to flush the system of fertilizer. The point of injection of the concentrated fertilizer solution should be after the filter, in order to prevent damage to the filter by corrosion. The fertilizer solution itself should be filtered before the injection pump (Burt et al. 1995).

The second step in fertigation design is determining the amount of fertilizer needed and the rate of injection, which may change according to the plant growth stage. It is beyond the scope of this book to give specific fertilizer recommendations for different crops. The reader is referred to detailed treatments of this problem by Burt et al. (1995), Bar-Yosef (1998), Scaife and Bar-Josef (1995), and Van Goor et al. (1988). If the total amount of fertilizer in terms of kg N, P, K and microelements per hectare is known, the fertilizers to be used can be chosen. Many fertilizers in solid form are suitable for fertigation, provided they are completely soluble. Soluble dry fertilizer containing N, P and K in different combinations are also on the market. Liquid fertilizers with varying N, P and K contents are also available but these are more expensive and more bulky to handle. It is possible to prepare solutions from mixtures of relatively inexpensive fertilizers depending on their solubility and compatibility. The main fertilizers used, their nutrient content, solubility and compatibility are given

Table 4.6. Fertilizers suitable for fertigation (after Burt et al. 1995; Lupin et al. 1996; Bar-Yosef, 1998)

Name	Chemical form	$N\text{-}P_2O_5\text{-}K_2O$ Content (%)	Solubility g/l at 20°C	Remarks
AmmoniumNitrate	NH_4NO_3	34-0-0	1830	Incompatible with acids
AmmoniumSulfate	$(NH_4)_2SO_4$	21-0-0	760	Clogging with hard water
Urea	$CO(NH_2)_2$	46-0-0	1100	
Urane solution	$CO(NH_2)_2N$ H_4NO_3	32-0-0	High	Incompatible with $Ca(NO_3)$
Monoammonium Phosphate	$NH_4\,H_2PO_4$	12-61-0	282	Not to be used with hard water (containing Ca)
Diammonium Phosphate	$(NH_4)_2HP_2$ O_5	18-46-0	575	Contains Phophorus at high solubility
PotassiumChloride	KCl	0-0-60	347	Chloride toxic for some crops, Cheapest K source
PotassiumNitrate	KNO_3	13-0-44	316	Expensive, high Nitrate
PotassiumSulfate	K_2SO_4	0-0-50	110	Excellent source of sulfur, Clogging with hard water
MonoPotassium Phosphate	KH_2PO_4	0-52-34	230	
Phosphoric acid	H_3PO_4	0-52-0	457	Incompatible with Calcium

in Table 4.6. The solubility of most fertilizers decreases at lower temperatures, which may cause precipitation in the cold season.

The rate of injection of fertilizer solution into the system q_c in l/h can be calculated according to Keller and Bliesner (1990)

$$q_c = F_r\, A/ct_r\, T, \tag{4.30}$$

where F_r is the fertilizer application rate per irrigation (kg/ha), A is the area irrigated (ha), c is the concentration of the nutrient in the fertilizer solution (kg/l), T is the duration of irrigation (h) and t_r is the ratio between fertilizing time and irrigation time. The exact fertilizer application rates to be applied to certain crops according to the actual uptake as it changes according to the plant growth rate are given by Bar-Yosef (1998) and Scaife and Bar-Yosef (1995). Usually, this kind of detailed information is neither available nor easy to implement, since the composition of the fertilizer solution and the fertilizer application rates must be changed continuously. Soil solution and plant tissue tests can be used to monitor the efficiency of the fertigation procedure. An excellent review of these methods and some general guidelines for the fertigation of several crops are given by Burt et al. (1995).

4.8
Subsurface Drip Design

Reasons for the increased use of subsurface drip systems are operational advantages: less labor involved, less soil evaporation, better trafficability and weed control, underground discharge of treated effluent water reduces possibility for pathogen exposure, and in some cases an increase in water use efficiency (Phene 1995). A subsurface system is similar in design to surface systems, but with some unique features. The main reason for installing a subsurface system is its sustainability. Therefore, high quality laterals and emitters should be chosen that can last a decade or longer. The design and layout of the system should be meticulous, since leaks in the system are difficult to locate and costly to repair. Primarily, clogging should be prevented at all cost, by

1. Adequate filtration and frequent cleaning of the check filters,
2. Installation of vacuum relief valves or vacuum breakers at the head of the system and at high elevations along the laterals. If the system closes, a vacuum may develop which may cause the sucking of sand or fine soil particles into some emitters, causing clogging.
3. Frequent flushing of the system, by installing automatic flushing valves at the end of each lateral or by connecting the laterals into a buried flushing submain with automatic or manual flushing.
4. Preventing root intrusion into the emitters by using a herbicide (treflane), dilute acid or by irrigation at high frequency, which results in permanent saturation of the soil in the close vicinity of the emitter.

Subsurface laterals in field or vegetable crops should be installed at a depth of 40–50 cm in order to prevent damage by tillage implements. In orchards along the tree rows a depth of 25 cm is sufficient.

4.9
Drip Irrigation in Greenhouses

Drip irrigation is an appropriate irrigation method for greenhouses, since it enables precise application of water and fertilizers. The high-value crops produced in greenhouses (flowers, vegetables, potting plants) have some special requirements of the irrigation system so that a high degree of uniformity in application is ensured. Since water is applied frequently, in most cases several times daily, rapid filling and emptying of the laterals is necessary. Devices that prevent back-flow of water and keep the laterals filled when the valves are shut (tube non-leakage) should be installed. Low-flow emitters (1–2 l/h) at close spacing (15–40 cm) are suitable for the irrigation of beds of soil and substrates. Pressure compensation of the emitters offers additional precision in water application. Individual plants or pots can be irrigated by non-leakage on-line pressure-compensated emitters, which "lock" themselves after each application. Some of these emitters have multiple outlets, enabling simultaneous irrigation of several plants. Several of these devices are shown in Fig. 4.12. The greenhouse irrigation and fertigation systems require automatic control and monitoring, as described in Section 2.9.

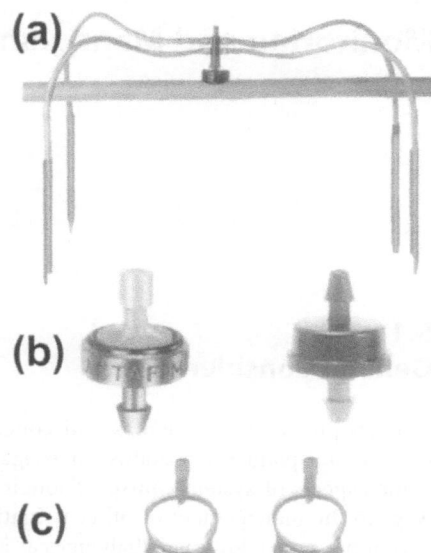

Fig. 4.12. Drip irrigation devices for green-house application: **a** compensated non-leakage (C.N.L.) dripper, **b** P.C.J. on-line drippers, **c** microtube adaptor for for RAM pressure compensated dripperline (By courtesy of NETAFIM Irrigation Equipment and Drip Systems)

Monitoring and Management of Drip Systems

5.1
General Considerations

This chapter addresses issues and concerns related to information on soil water, climate and plant water status for irrigation and fertigation scheduling, as well as some aspects of system control. Though the general principles are similar in many ways to the management of other irrigation methods, the management of drip irrigation presents additional challenges as a result of high irrigation frequency, limited soil wetting, and technical skills required for proper system operation and maintenance. Management at high efficiencies common under drip irrigation invariably means operating with limited soil water and nutrient storage (to reduce drainage losses). Such limited capacitance leaves a small margin for error; hence the success of drip irrigation management hinges on precise information on crop water use, soil water status, and system delivery to ensure correct timing and amounts of water and nutrient application.

In addition to the small margin for error, drip systems are susceptible to emitter clogging; therefore, proper operational and preventative maintenance is imperative. Many modern drip irrigation systems are equipped with such feedback and control elements as water meters, pressure gauges and controllers. These elements are not only used for routine operations (e.g., dosing and measuring amounts of water applied), but may also be used to provide an updated picture of system performance (e.g., monitoring changes in flow rates indicating emitter clogging).

5.2
Irrigation Scheduling and Water Balance

Irrigation scheduling is commonly defined as determining the timing and amount of irrigation water to be applied to a given crop area. The decision will vary depending on the objectives of the irrigator: (1) maximum economic return; (2) maximum yield per unit area; or (3) maximum yield per unit water. Each objective requires an optimization of a different strategy or measure of performance. These might "measure" the net benefits per unit area (or unit water), a crop production function, minimum deviations from a certain optimal soil water status, or minimum deviations from certain evapotranspiration values.

A growing plant takes up soil water by its roots and transmits the water to the atmosphere through the leaves in a physical-biological process called *transpiration* (T). Water vapor may also reach the atmosphere by direct evaporation from the soil in the process of *evaporation* (E). These two independent processes are difficult to separate under partial plant cover in cropped fields and are thus often combined and treated as a single process called *evapotranspiration* (ET). Transpiration rates (T) are dependent upon external climatic conditions (the evaporative demand of the atmosphere), as well as on the availability of soil water and the ability of the soil to transmit water at sufficient rates. Studies have shown that plant transpiration rates are affected by the water status in their rooting zone as the availability between field capacity and wilting point gradually decreases.

The objective of irrigation in its broader context is the replenishment of soil water lost to ET and the maintenance of favorable growth conditions in plant rooting zones. The determination of irrigation amounts to be replenished may be based on climatic measurements of ET, or by direct measurements of changes in soil water storage. Irrigation timing is determined such that only a certain amount of depletion is allowed before replenishment is scheduled. The primary reason for maintaining certain levels of soil water is because not all the stored water in the soil is equally available for plant use. As depletion progresses, extraction of the remaining water by plant roots becomes progressively more difficult and the plant may experience water stress caused by reduced transpiration, which, in turn, leads to a potential reduction in growth and yields.

The challenge in irrigation scheduling is the estimation of the depletion for a given time period. Crop consumptive use is not constant and varies with the growth stage and climatic conditions. The framework most often used is the soil water balance whereby all inputs and outputs are estimates to determine the amount of irrigation required. The basic equation describing the soil water balance is:

$$I = ET + DR + RO - \Delta W - P, \tag{5.1}$$

where I is irrigation, ET is evapotranspiration (soil evaporation + plant transpiration), DR is drainage and deep percolation, RO is surface runoff, ΔW is change in water storage within the profile (soil water depletion), and P is precipitation. W is defined as the equivalent depth of water stored in the soil profile under consideration during period i, and $\Delta W = (W_{initial} - W_{final})$. The quantities in this equation are all associated with a given specific time interval (the equation can represent a daily, weekly, or annual water balance in a soil profile). The convention often used is that inputs to the profile are taken as positive, and outputs have a negative sign. The concept is based on conservation of mass (water) and is similar to the familiar exercise of balancing inputs and outlays from a checking account. In many arid regions, the components P, RO, and DR may be neglected. Thus Eq. (5.1) simplifies to $ET = I + \Delta W$, reflecting the fact that depletion equals evapotranspiration in the absence of irrigation (i.e., during an irrigation interval starting after an irrigation, hence I = 0). If, on the other hand, the water balance is calculated for an interval starting prior to an irrigation, $\Delta W \approx 0$ (approximately for many regular intervals), and I = ET. These simple relationships provide the motivation for soil water content measurements and estimation of ET for irrigation scheduling.

This simple picture is somewhat marred by the fact that all quantities in Eq. (5.1) are expressed as equivalent depths per land area (actually fluxes, or volumes of water per unit surface area per time interval), whereas the partial wetting and the multidimensional flow and uptake under drip are better described by volumes. In practice, these limitations of the one dimensional description are circumvented by considering both irrigation and ET per unit area, and integrating soil water measurements to represent the equivalent storage in a soil "profile".

5.2.1
Irrigation Amounts and System Capacity

Two of the most important factors for design and irrigation management are: (1) the peak water requirement, and (2) the seasonal water use. The peak water requirement is the irrigation application rate, or for a pre-set irrigation interval, the irrigation amount, needed to meet the largest crop water requirement. In most circumstances daily crop water use does not exceed 10 mm/day (expressed per unit area of the irrigated field), and a more reasonable value for design purposes is 7 mm/day.

For a peak water requirement the minimum system capacity is calculated according to:

$$Q_{min} = \frac{ET_{peak}A}{T_iEa},$$
(5.2)

where Q_{min} is the volumetric flow rate (e.g., m^3/day or m^3/h) for the set or an irrigated plot, A is the irrigated area (m^2), T_i is irrigation "on" time (fraction of a day, or h/day), ET_{peak} is peak water requirement (m/day, or mm/day with the proper unit conversion), and Ea is irrigation efficiency for drip systems at about 90% (or 0.9). The actual system capacity should exceed Q_{min} to meet irrigation demand and allow flexibility in irrigation scheduling.

5.2.2
Effect of Time of Day

For drip systems irrigated at a regular interval, irrigation timing with respect to time of day plays an important role in the amplitude of soil water content and matric potential diurnal cycle, and affect the potential for deep percolation. For daytime application, crop ET takes up the water as it is being applied and only a portion goes to storage. However, when water is applied at night it must be either stored in the profile or lost to deep percolation. An experimental study conducted by Abbott and Koon (1992) demonstrated the development of two different soil water regimes under sugar cane irrigated at night vs. daytime. The main findings are depicted in Fig. 5.1 clearly showing the larger and deeper extent of the wetted soil volume for day-irrigated sugar cane relative to night-irrigated treatment. At first glance, the result appears somewhat counterintuitive and inconsistent with basic soil physics principles (one would expect larger drainage losses for night-time irrigation due to redistribution in the absence of plant water uptake). However, the dryer soil profile at the onset of night-time irrigation reduces the hydraulic conductivity relative to the daytime irrigated profile whereas water redistributed during the night makes the profile "uniformly wetter" on average. The primary result here is the more controlled oscillations of soil water matric potential for daytime irrigation relative to the large changes for night-time irrigation.

In summary, irrigation timing relative to plant uptake and redistribution time may

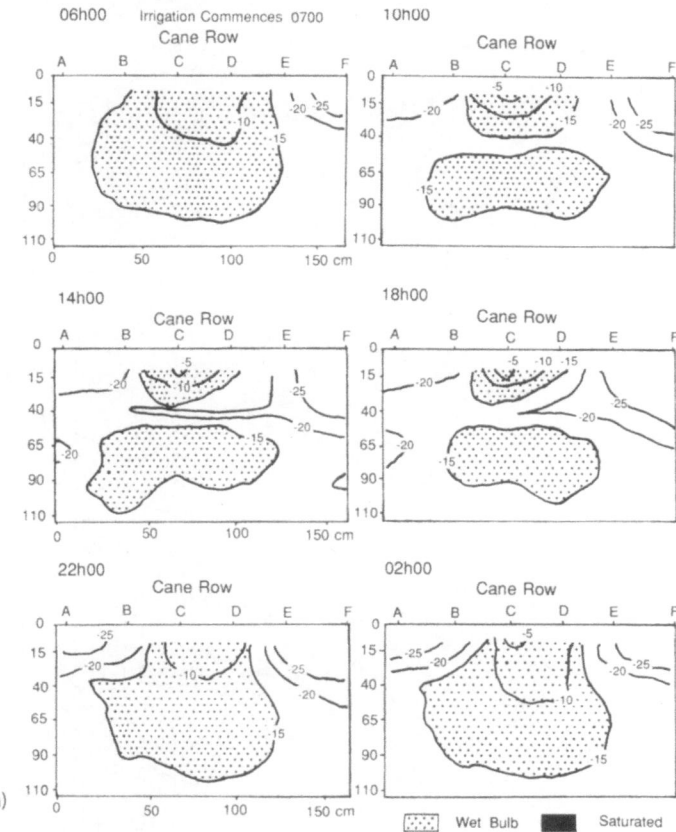

Fig. 5.1. Two-dimensional distribution of total soil water potential (kPa) for a 24-h period starting at 6 a.m. for irrigation of sugarcane **a** by day and **b** by night (Abbott and Koon 1992)

play an important role in determining deep percolation losses especially in coarse-textured soils, and for daily to bi-daily intervals (for longer intervals irrigation time may take the entire day). It also plays an important role in the determination of diurnal cycle in matric potential.

5.2.3
Evapotranspiration and Crop Coefficients

A limited discussion of methods used for estimating evapotranspiration from climatic measurements is given in Section 5.3.2. In the following, we present basic energy balance, general climatic methods for ET estimation, and introduce the concepts of potential and actual ET and crop coefficients used for determining irrigation amounts.

Briefly, the net solar radiation (R_N) impinging on the earth's surface may be partitioned into one of several components. Part of the energy may be transformed into

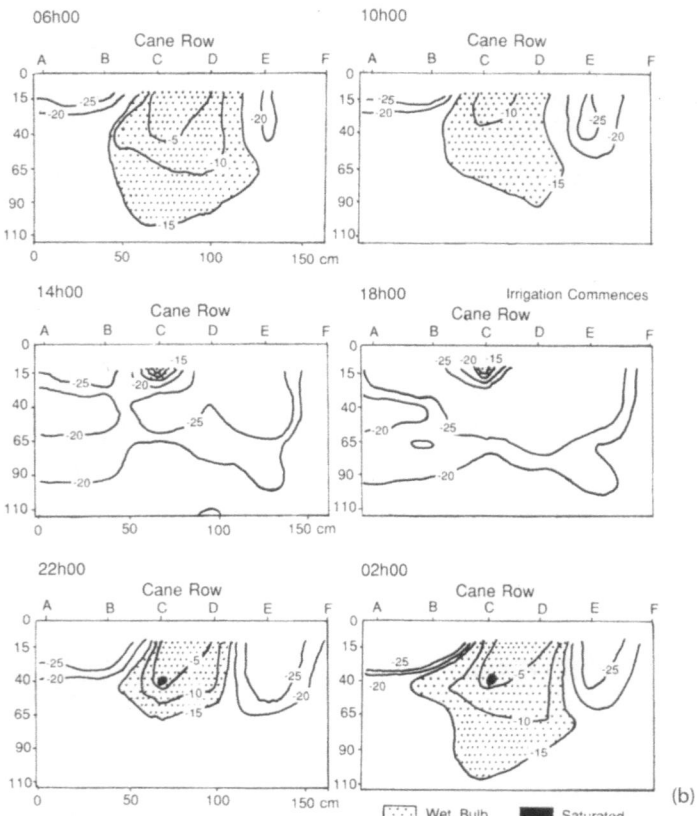

Fig. 5.1. *Continued*

heat that warms the soil, plants and the atmosphere; another part may be used by plants for photosynthesis. A major part of R_N, however, is used for evapotranspiration (ET). The energy balance on a field-scale surface is given by:

$$R_N \downarrow = H \uparrow + LEt \uparrow + G \downarrow, \qquad (5.3)$$

where H is the energy utilized in heating the air (*sensible heat flux*), G is the energy utilized in heating the soil (*soil heat flux*), L is the latent heat of vaporization (2.449 MJ/kg or 585 cal/gr), and LEt is the *latent heat flux*. Et has units of length, while multiplication by the latent heat of vaporization gives LEt units of W m^{-2}. The convention of signs [shown by the arrows in eq. (5.3)]) indicates the direction of energy flow (relative to the soil surface), which is considered positive.

One method for estimating ET, the water balance method, was discussed in the previous section. It is based on measurements of water balance components, e.g., rainfall, irrigation, runoff, changes in soil water storage, and deep percolation, to estimate the unknown value of ET. The method is relatively expensive and labor intensive, and is sensitive to spatial variations in soil properties and plant cover. The energy balance equation and equations describing the transport of water vapor into the atmosphere offer an alternative to estimating ET from measurement of climatological attributes (incoming radiation energy, temperature, etc.) and estimation of vapor and heat transfer coefficients into the atmosphere.

Climatically-based methods for ET estimation may be classified as: (1) Aerodynamic profile methods – based on estimation of vapor and heat fluxes using their respective transport coefficients and their vertical gradients; (2) energy balance methods – directly measuring or estimating all attributes other than ET (similar to the water balance approach); (3) combination methods – combination of the above approaches; (4) methods based on turbulence measurements and on variance dissipation; and (5) empirical methods – based on empirical relationships between simple-to-obtain attributes (e.g., air temperature, solar radiation) and ET, and usually applicable only to monthly or seasonal ET estimates.

Plants can influence ET in several ways, including stomatal closure, partial ground cover, and changes in rooting depths. Thus it is difficult to factor these influences into general ET estimation. To circumvent this problem, the concept of potential ET (or ETp) was introduced; it is defined as the highest value of ET possible based on available energy [eq. (5.1)]. A related concept is the reference crop ET (ETr) defined as the maximal ET obtainable from well-watered short crop (Jensen 1973; Hanks 1983). Most standard methods provide estimates of ETp or ETr. A typical seasonal pattern of ETp measured near Bishop, California, is depicted in Fig. 5.2 (Or and Groeneveld 1994). This figure also shows the relatively low standard deviations in multi-year daily values of ETp in that region.

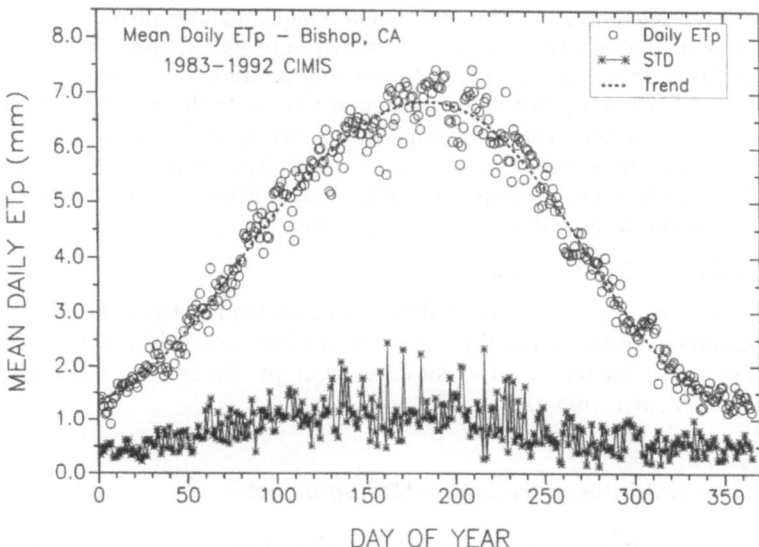

Fig. 5.2. Mean daily potential evapotranspiration and its standard deviation estimated by the modified Penman equation from climatic data collected at Bishop, California (Or and Groeneveld 1994)

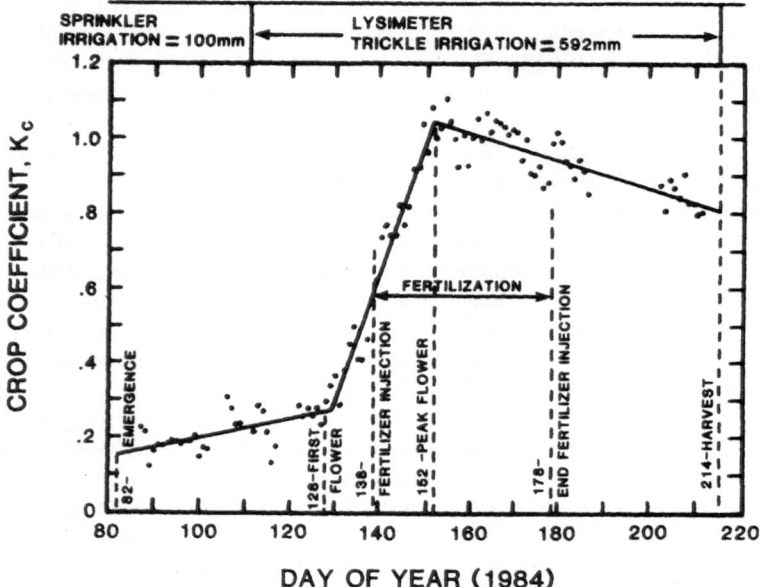

Fig. 5.3. Crop coefficient curve for a sub-surface drip-irrigated tomato crop (Phene, 1986)

Irrigation scheduling should be based on actual ET (ETa) rather than on the potential (or reference) ET. However, detailed modeling and prediction of transpiration rates are impractical due to the complexity of processes involved in plant growth and function. Practical estimation of plant transpiration rates is often based on relating actual rates to a potential or maximum rate controlled by atmospheric conditions (e.g., ETp or ETr). This is accomplished by using empirical matching factors, known as crop coefficients (Kc), which may be time-, species-, or site- dependent (Fig. 5.3). The practical use of crop coefficients relates actual ET (ETa) to ETp, using crop coefficients as growth stage matching factors:

$$ETa(t) = ETp(t) \cdot Kc(t). \tag{5.4}$$

Numerous refinements are available for these simple relationships (e.g., the incorporation of a water availability factor). Doorenbos and Pruitt (1977) provide a comprehensive source for Kc values for different crops . An updated version is forthcoming (R.Allen, pers. comm).

5.2.4
Threshold-Based Irrigation Management

In cases where constant water supply is available and capacity is not overly constrained, pre-defined soil water status in a crop root zone could be maintained either by means of automatic irrigation or by determining irrigation timing as part of a scheduling scheme. Drip irrigation is particularly suitable for this type of irrigation management as it offers some advantages over irrigation timing based on water balance or ET calculations. Minimizing deviations from a certain range of soil water conditions known to be favorable for crop development becomes the objective of irrigation scheduling. Soil water sensors could be used to provide automatic input to an irrigation controller and activate irrigation whenever a pre-set threshold is exceeded (e.g., Cary and Fisher 1983; Phene and Howel 1984).

The use of index tensiometers to determine irrigation timing as part of irrigation scheduling has been recommended since the early 1950s (Taylor 1965; Campbell and Campbell 1982; Bell et al. 1990). Most of these studies focused on defining ranges of matric potentials for optimal crop yields (Taylor 1965; Phene and Beale 1976; Phene and Sanders 1976; Phene and Howell 1984). As Hodnett et al. (1990) point out, the drawbacks of such guidelines are their empirical bases and lack of generality, i.e., they are valid only for the specific soil, crop, and positions for which the results were obtained.

The performance of a sensor-based drip irrigation scheduling scheme is dependent upon both sensor location and its operational threshold (the sensor value for onset of irrigation). Much effort went into developing agronomically-desirable thresholds, whereas the question of sensor placement within non-uniformly-wetted soil volumes was often neglected (Hodnett et al. 1990). Many of the existing recommendations for sensor placement are qualitative and empirically-based (Pogue and Pooley 1985). For example: (1) sensors should be placed at the top and bottom of the active rooting zone (Haise and Hagan 1967); or (2) sensors should be placed relatively close to the dripper (Phene and Howell 1984, Levin et al. 1985). Several studies attempted to define the number of measurements necessary to obtain an accurate estimate for a threshold (Doorenbos and Kassam 1979; Hendrickx and Wierenga 1990); others evaluated the impact of spatial and temporal variability in soil properties on the variability of point measurements (Hendrickx and Wierenga 1990; Or 1996).

Recent studies by Coelho and Or (1996) attempted to develop a more rigorous framework for sensor placement based on comprehensive considerations of soil water dynamics within the crop root zone (using a modeling approach). A certain threshold-location would result in a certain irrigation interval and associated soil water dynamics. Conversely, any pre-set irrigation interval (assuming replenishment based on ET), may be characterized (and controlled) by a large number of threshold-location combinations. In other words, a choice of threshold-location sets a certain irrigation interval and typical soil water regime in the root zone (assuming that ET does not vary much during a typical irrigation cycle). For example, placing the sensor very close to the emitter, and selecting a relatively "wet" threshold, could induce multiple irrigations in the course of a single day (see Fig. 5.4). A potentially useful aspect of the Coelho and Or (1996) study was the definition of soil volumes unsuitable for tensiometric monitoring (when a predefined irrigation interval is the primary decision variable) due to the exceedingly large matric potential fluctuations. Similar guidelines may be developed for suitable locations of other types of sensors based on their range of operation and sensitivity in relation to the expected soil water regime in the wetted root zone.

In summary, for automatic drip irrigation based on soil water sensors, the joint effect of a threshold value and a particular sensor location (relative to the dripper) define an "effective" reservoir size of available soil water for plant uptake. Hence, when

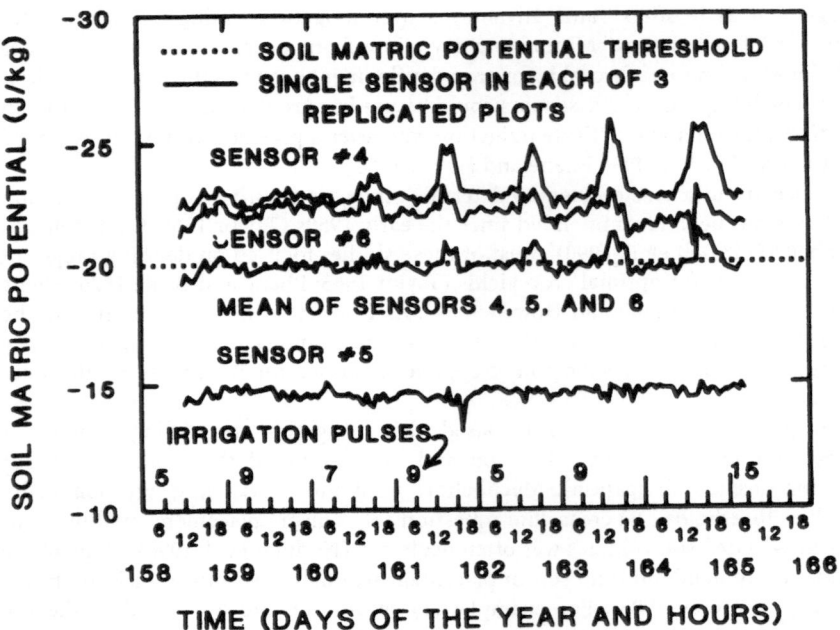

Fig. 5.4. Fluctuations of soil matric potentials at a soil depth of 45 cm measured by sensors for a constant threshold of −20 kPa and the number of pulse irrigations applied automatically each day over an eight-day period (Phene and Howell 1984)

evapotranspiration is relatively constant, an "effective reservoir" size, as defined by a sensor threshold and location, also results in a nearly regular irrigation interval. An example of automatic irrigation scheduling based on soil matric potential sensor is depicted in Fig. 5.4 (Phene and Howell 1984).

5.3
Monitoring Methods, Sensors, and Irrigation Guidelines

The depth of water D_I multiplied by the area to be irrigated gives us the the volume of water to be applied for a certain irrigation. However, the biggest problem is how to obtain a reliable estimate of ETa, the actual water requirement of the crop. ETa is determined by soil water status, plant water status and climatic factors. Each of these factors can be measured by sensors and used as feedback to automatically schedule irrigation.

5.3.1
Soil-Based Sensors for Monitoring Soil Water Status

Soil water status can be measured by means of tensiometers, which measure soil water potential in the limited range of 0–0.8 bar. They can be monitored with pressure transducers and used as input to an irrigation controller (Meron et al.

1995). These are essentially point measurements and data from several instuments have to be integrated in order to obtain a reliable estimate. Dielectric measurement of soil water content by time domain reflectometry or by capacitance measurements have developed recently but are not economically justified for use in commercial fields. Table 5.1 gives an overview of the methods for soil water status monitoring.

Soil water status is determined either by the amount of water stored in the soil (often measured as the volumetric water content), or by the energy state of soil water (i.e., the matric potential that reflects the energy by water retained in the soil). The linkage between water content and matric potential is described by a soil characteristic or retention curve. These relationships are dependent on soil type and are highly nonlinear. Typical retention curves for different textural groups are depicted in Fig. 5.5 (Leij et al. 1999).

Typically, higher (less negative) potentials are favorable for plant growth and are conducive for water uptake (because plant roots need to overcome only low water

Table 5.1. Presently available methods for soil water status monitoring in irrigation scheduling (Coelho and Or 1996)

Method	Attribute	Range of Operation	Main Advantages	Main limitations
Gravimetry	θ_g	Air-dry to Saturation	Simple, inexpensive, accurate	Non-repeatable, time-consuming, requires bulk density for θ_v and irrigation amounts
Neutron scattering	θ_v	Air-dry to Saturation	Rapid, repetitive, non-destructive measurements	High cost, radiation hazard, unreliable near soil surface, requires site specific calibration
Electrical Resistance	H	$h < -1\,\mathrm{m}$	Low cost, simple, repetitive, non-destructive, multiplexing	Unreliable in coarse-textured soils, sensor variability, inaccurate in high water content, requires calibration, unreliable in high salinity
Heat Dissipation	H	$h > -20\,\mathrm{m}$	Repetitive, non-destructive, multiplexing	Considerable cost, calibration needed, sensor variability
TDR	θ_v (and EC)	Air-dry to saturation	Repetitive, non-destructive, high accuracy, modest calibration, high spatial-temporal resolution, multiplexing	High cost, limited in salt affected soils, calibration for organic soils
Tensiometer	h	$h > -8\,\mathrm{m}$	Simple, reliable, repetitive, (multiplexing with transducers)	Frequent servicing, considerable cost, limited range

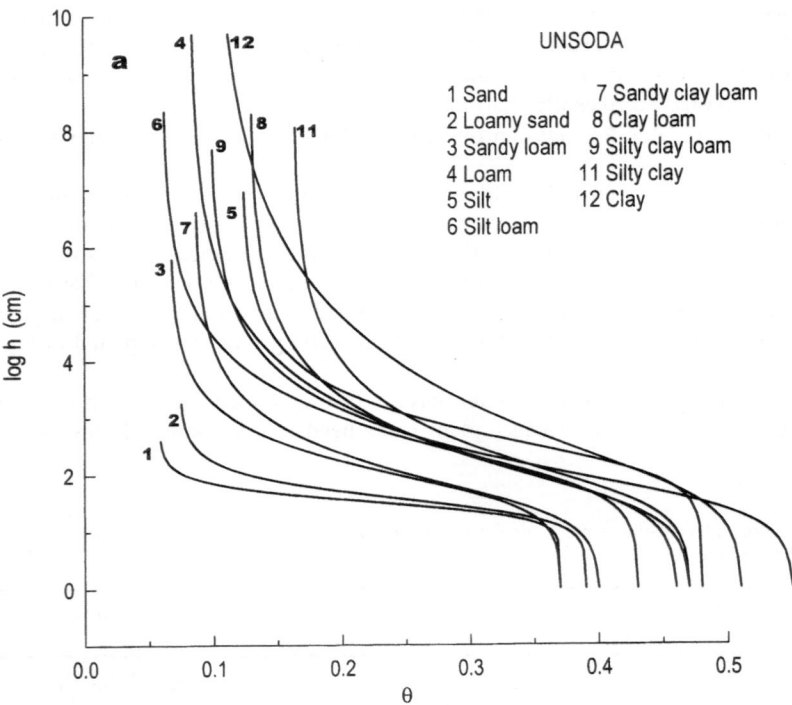

Fig. 5.5. Typical shapes of soil-characteristic curves for different soil types (Leij et al. 1998)

energy retention by soil). However, as saturation is approached, aeration problems may develop due to the low gaseous diffusion coefficient which is determined by fewer and fewer air-filled pores. Other important reasons for avoiding matric potential values near complete saturation, are the increased hydraulic conductivity, the increased risk of large amounts of deep percolation, and the leaching of nutrients. At potentials near saturation it becomes very difficult to maintain a target potential value, and the irrigation system becomes extremely inefficient.

Defining favorable conditions in terms of matric potential is more general than using water content values because of the large variations between different soils. Reasonable values for matric potential (near the soil water content at field capacity) range from −10 to −20 kPa for sandy soils, and −20 to −40 kPa for fine-textured soils. These values are for sensors placed near an emitter within the frequently wetted zone. The following tables provide typical values of soil water content and matric potential to be maintained for optimal yield of different crops. The data presented in Table 5.2 are based on the concept of the stress day index (SDI). This index attempts to quantify the amount of stress imposed on a crop during different growth stages (and different susceptibilities), and provides general guidelines for irrigation timing for different crops. An even more important consideration is that the threshold value varies for different growth stages.

Table 5.2. Practical guide to irrigation timing using the stress day index SDI method (Hiller and Howell 1983)

Growth Stage	Yield Reduction if water-stressed (%)	To avoid stress Irrigate when soil Moisture depleted reaches: (%)
Grain sorghum		
Vegetative (6- to 8- leaf stage)	25	65
Boot to heading	36	45
Heading to soft dough	45	35
After soft dough	25	65
Cotton		
Prior to flowering	00	–
Early flowering	21	60
Peak flowering	32	35
Late flowering	20	60
Soybeans		
Vegetative	12	80
Early-to-peak flowering	24	45
Late flowering, early pod development	35	30
Late pod development to maturity	13	80
Corn		
Vegetative	25	65

Another table with "recommended" threshold values (in this case in terms of matric potential) was compiled by Taylor and Ashcroft 1972). Although the sources for the data in Table 5.3 are from 1950 to 1965, the values are reasonable, considering the wide variations in soil types and management.

Some of the common devices and sensors used for soil water status monitoring (grouped by sensors for water content measurements) are the neutron scattering method, time domain reflectometry, and several new capacitive and frequency domain sensors. Methods for measurement (or inference) of matric potential include tensiometer, psychrometer, heat dissipation sensors, and resistance blocks. A distinction between methods based on manual measurements and sensors capable of providing a constant output, is important for proper consideration in an automatic irrigation scheme.

Neutron Scattering (Volumetric Water Content, Manual). This nondestructive method for repetitive field measurement of volumetric water content is based on the propensity of hydrogen nuclei to slow (thermalize) high- energy fast neutrons. A typical neutron moisture meter consists of: (1) a probe containing a *radioactive source* that emits high energy (2–4 MeV) fast (1600 km/s) neutrons, as well as a *detector* of slow neutrons; (2) a *scaler* to electronically monitor the flux of slow neutrons; and, optionally, (3) a data logger to facilitate storage and retrieval of data (Fig. 5.6). When the probe is lowered into an access tube, fast neutrons are emitted radially into the

Table 5.3. Matric potentials at which water should be applied for maximum yields of various crops grown in deep, well-drained soil that is fertilized and otherwise managed for maximum production

Crop	Matric Potential (J/kg or centibars)	Crop	Matric Potential (J/kg or centibars)
VEGETATIVE CROPS		*FRUIT CROPS*	
Alfalfa	−150	Lemons	−40
Beans (snap and Lima)	−200 to −75	Oranges	−20 to −100
Cabbage	−70 to −60	Deciduous fruit	−50 to −80
Canning peas	−50 to −30	Avocados	−50
Celery	−30 to −20	Grapes	
Grass	−100 to −30	Early season	−40 to −50
Lettuce	−60 to −40	During maturity	<−100
Tobacco	−80 to −30	Strawberries	−20 to −30
Sugar Cane	−50 to −15	Cantaloupe	−30 to −40
Sweet corn	−100 to −50	Tomatoes	−80 to −150
Turf Grass	−36 to −24	Bananas	−30 to −150
Broccoli			
Early	−55 to −45	*SEED CROPS*	
After budding	−70 to −60	Alfalfa	
Cauliflower	−70 to −60	Prior to bloom	−200
		During bloom	−400 to −800
		During ripening	−800 to −1500
ROOT CROPS		Carrots during seed year	−400 to −600
Onions:		Onions during seed year	
Early growth	−55 to −45	At 7 cm depth	−400 to −600
Bulbing time	−65 to −55	At 15 cm depth	−150
Carrots	−65 to −55	Lettuce during productive phase	−300
Sugar beets	−60 to −40		
Potatoes	−50 to −30		
		Coffee	Requires short Periods of low potential to break bud dormancy, followed by high water potential
GRAIN CROPS			
Corn			
Vegetative period	−50		
During ripening	−1200 to −800		
Small grains			
Vegetative period	−50 to −40		
During ripening	−1200 to −800		

soil where they collide with various atomic nuclei. Collisions with most nuclei are virtually elastic, i.e., with only a minor loss of kinetic energy by the fast neutrons. Collisions with hydrogen nuclei, which have a similar mass to neutrons, cause a significant loss of kinetic energy and slow down the fast neutrons. The slow neutrons rapidly form a "cloud" of nearly constant density near the probe, where the flux of the slow neutrons is measured by the *detector*. The relative number of slow neutrons is therefore proportional to the amount of hydrogen nuclei in the surrounding soil. The primary source of hydrogen in soil is water; other sources of hydrogen in a given soil are assumed constant and are accounted for during calibration. Although several non-

Fig. 5.6. A schematic illustration of a neutron-probe device for measuring soil water content in a wet and in a dry soil

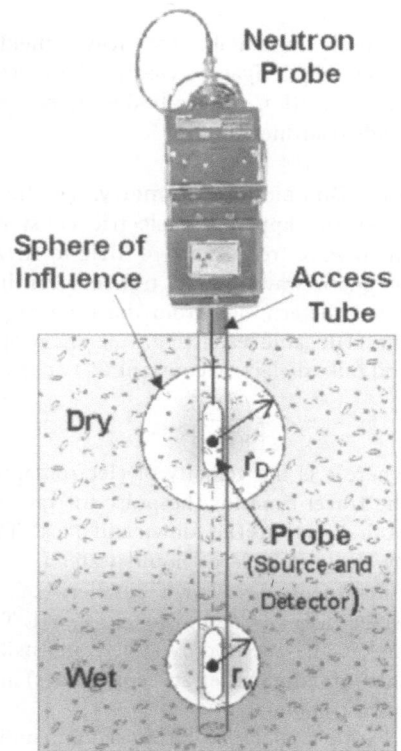

hydrogen substances, which may be present in trace amounts in some soils, may also thermalize fast neutrons, these also may be effectively compensated for through soil-specific calibration. Calibration of the neutron probe is required to account for background hydrogen sources and other local effects (e.g., soil bulk density), and is conveniently achieved by simultaneous measurements of soil water content and neutron probe counts. The calibration curve is usually linear and relates volumetric water content to slow neutron counts or count ratio (CR):

$$\theta_v = a + b(CR), \tag{5.5}$$

where CR is the ratio of slow neutron counts at a specific location in the soil to a standard count obtained with the probe in its shield. For many soils the calibration relationship is approximately the same. Use of the count ratio rather than raw slow neutron counts compensates for the slow decay of the radioactive source over time.

The sphere of influence about the radiation source varies between about 15 cm (wet soil) to perhaps 70 cm (very dry soil), depending on how far fast neutrons must travel in order to collide with a requisite number of hydrogen nuclei. An approximate equation for the radius of influence (r, in cm) as a function of soil wetness is: $r = 15\,(\theta_v)^{-1/3}$.

Thus, the neutron scattering method is unsuitable for measurement near the soil surface because a portion of the neutrons may escape the soil. Similarly, a neutron probe is not suitable for measurements near the fringes of the wetted soil volume under drip irrigation. Despite these two constraints to the use of a neutron probe for drip irrigation, the method is widely used in research and commercial drip-irrigated fields (Gardner 1986).

Time Domain Reflectometry. The time domain reflectometry (TDR) method measures the apparent dielectric constant of the soil surrounding a waveguide, at microwave frequencies of MHz to GHz. The propagation velocity (v) of an electromagnetic wave along a transmission line (waveguide) of length L embedded in the soil is determined from the time response of the system to a pulse generated by the TDR cable tester (Fig. 5.7). The propagation velocity (v = 2 L/t) is a function of the soil bulk dielectric constant (ε_b) according to

$$\varepsilon_b = \left(\frac{c}{v}\right)^2 = \left(\frac{ct}{2L}\right)^2 , \tag{5.6}$$

where c is the velocity of electromagnetic waves in vacuum (3×10^8 m/s), and t is the travel time for the pulse to traverse the length of the embedded waveguide in both directions (i.e., down and back). The soil bulk dielectric constant (ε_b) is governed by the dielectric of liquid water $\varepsilon_w = 81$, as the dielectric constants of other soil constituents are much smaller, e.g., for soil minerals $\varepsilon_s = 3$ to 5, for frozen water (ice) $\varepsilon_I = 4$, and for air $\varepsilon_a = 1$. This large disparity of the dielectric constants makes the method relatively insensitive to soil composition and texture (other than organic matter and some clays) and thus is a good method for liquid soil water measurement.

Topp et al. (1980) proposed an empirical method for the relationships between ε_b and volumetric soil water content (θ_v) based on a third-order polynomial fitted to measured ε_b and θ_v for multiple soils:

$$\theta_v = -5.3 \times 10^{-2} + 2.92 \times 10^{-2}\,\varepsilon_b - 5.5 \times 10^{-4}\varepsilon_b^2 + 4.3 \times 10^{-6}\varepsilon_b^3 \tag{5.7}$$

Other expressions based on dielectric mixing models are also frequently used.

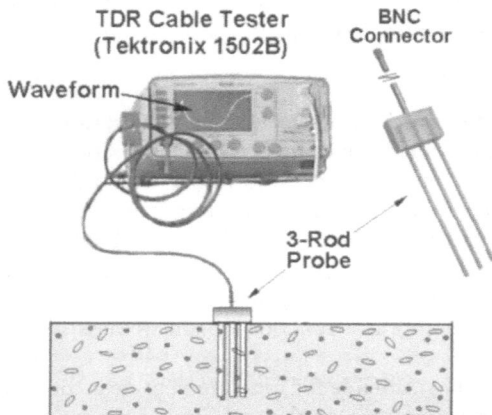

Fig. 5.7. Time domain reflectometry system showing a cable tester with a 3-rod probe inserted vertically into the soil

The relatively small probes used with TDR, the potential for continuous output, and high accuracy could make this method ideal for drip irrigation, provided the high cost of the instruments drops to economical levels. To summarize, the main advantages of TDR over other methods for repetitive soil water content measurement (e.g., neutron probe) are: (1) superior accuracy to within 1 or 2% of volumetric water content; (2) calibration requirements are minimal (in many cases soil-specific calibration is not needed); (3) radiation hazards associated with neutron probe or gamma-attenuation techniques are averted; (4) excellent spatial and temporal resolution are attained; (5) measurements are simple to obtain, and (6) the method is capable of providing continuous soil water measurements through automation and multiplexing. Limitations of the TDR method include (1) relatively high equipment expense, (2) limited applicability under highly saline conditions due to signal attenuation, and (3) the soil-specific calibration required for soils with high clay or organic matter contents.

Frequency Domain and Other Capacitance Methods. Several new sensors and measurement methods are based on combinations of capacitive, reflective and frequency-shift principles, all of which are governed by soil dielectric properties. This trend appears highly promising for the development of accurate and cost-effective sensors for soil water content measurement and will likely dominate future developments in this area.

An example of such a stand-alone sensor is the *water content reflectometer* (Campbell Scientific Inc., Logan, UT) which provides an indirect measurement of soil volumetric water content based on changes in soil dielectric permittivity (Fig. 5.8). High speed electronic components are configured in an oscillator circuit which is connected to parallel rods acting as a waveguide. The rods are inserted in the monitored soil depth (typical rod length is 0.3 m). As soil water content changes, the resultant dielectric property causes a shift in the oscillation frequency of the circuit. A calibration relationship is established between the output frequency of the circuit and the soil volumetric water content. The time required for the actual measurement is less than 20 ms. The method is sensitive to soil electrical conductivity and an adjustment must be made to the calibration when soil solution conductivity exceeds $2\,dS\,m^{-1}$. Another

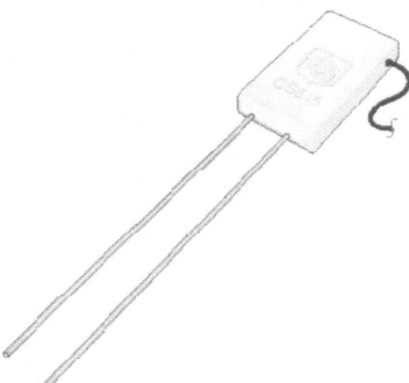

Fig. 5.8. Water content reflectometer sensor (Campbell Scientific Inc. Logan, Utah)

commercially-available stand-alone sensor based on frequency domain (FD) measurements is described in detail by Hilhorst and Dirksen (1994).

Tensiometer. A tensiometer consists of a porous cup (usually made of ceramic having very fine pores) connected to a vacuum gauge through a water-filled tube (Fig. 5.9). The porous cup is placed in intimate contact with the bulk soil at the depth of measurement. When the matric potential of the soil is lower (more negative) than the equivalent pressure inside the tensiometer cup, water moves from the tensiometer along a potential energy gradient to the soil through the saturated porous cup, thereby creating suction sensed by the gauge. Water flow into the soil continues until equilibrium is reached and the suction inside the tensiometer equals the soil matric potential (ψ_m). When the soil is wetted, flow may occur in the reverse direction, i.e., soil water enters the tensiometer until a new equilibrium is attained.

Electronic sensors called *pressure transducers* often replace the mechanical vacuum gauges. The transducers convert mechanical pressure into an electric signal that can be more easily and more precisely measured. In practice, pressure transducers can provide more accurate readings than other gauges, and in combination with data logging equipment, are able to supply continuous measurements of soil matric potential.

The tensiometer range is limited to suction values (absolute value of the matric potential) of less than 100 kPa (i.e., 1 bar or 10 m head of water) at sea level, and this value decreases proportionally with elevation gain. Thus other means are needed to measure or infer soil matric potential under drier conditions (as discussed in the section on sensor placement).

Electric Resistance. Changes in the electrical resistivity of porous materials with changes in their water content (and with soluble ionic constituents) have been used

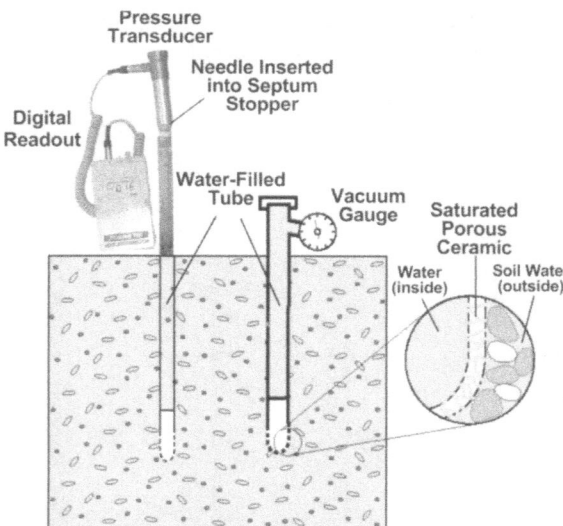

Fig. 5.9. Tensiometers with pressure transducer and vacuum gauge with details across porous ceramic cup

to develop simple and cheap sensors to infer soil water status. These sensors usually consist of concentric or flat electrodes embedded in a porous matrix and connected to lead wires for measurement of electrical resistance within the sensor's porous matrix. The commonly used term "gypsum block" arises from early models, which were, in fact, made of gypsum (Bouyoucous and Mick 1940), and from the practice of saturating the matrix of many sensors made from alternative materials with gypsum to buffer local soil ionic effects. The sensor is embedded in the soil and allowed to equilibrate with the soil solution. The matric potential of water in the sensor is determined from the measured electrical resistance through previously determined calibration of the sensor itself (i.e., electrical resistance vs. matric potential). Under equilibrium conditions the sensor matric potential is equal to the soil water matric potential; however, the sensor water content may be different from the soil. Hence, these measurements are often used to infer soil water matric potential from which the soil water content may be estimated, based on a known relationship between these quantities (Gardner 1984). With proper calibration for a particular soil the sensor could be used to infer soil water content directly (Kutilek and Nielsen 1994). The main advantages of electrical resistance sensors are their low cost and simple measurement requirements. Measurements may be obtained using a simple resistance meter, or more conveniently acquired automatically using a data logger. On the other hand, the usual requirement for specific calibration of each sensor and for each soil to obtain acceptable accuracy, and lack of sensitivity under wet conditions, render this measurement method appropriate mostly as a qualitative indicator of soil water status (Spaans and Baker 1992).

Heat Dissipation in a Rigid Porous Matrix. The rate of heat dissipation in a porous medium is dependent on the medium's specific heat capacity, thermal conductivity, and density. The heat capacity and thermal conductivity of a porous medium is affected by its water content, and hence related to its matric potential. Heat dissipation sensors contain in line or point source heating elements embedded in a rigid porous matrix with fixed pore space. The measurement is based on applying a heat pulse through application of a constant current through the heating element for a specified time period, and analyzing the temperature response measured by a thermocouple fixed at a known distance from the heating source (Phene et al. 1971; Bristow et al. 1994). Sensors are individually or uniformly calibrated in terms of heat dissipation vs. sensor wetness (i.e., matric potential). With the heat dissipation sensor buried in the soil, changes in soil matric potential result in a gradient between the soil and the porous matrix that induces a water flux between the two materials until a new equilibrium is established. The water flux changes the water content of the porous matrix which, in turn, changes the thermal conductivity and heat capacity of the sensor. In this manner the measured thermal response of the sensor may be related to soil wetness. A typical useful matric potential range for such sensors is −10 to −1000 kPa. An example of a line source heat dissipation sensor is depicted in Fig. 5.10.

In summary, this section provided an overview of soil water monitoring methods and sensors, as well as a range of optimal depletion levels (Table 5.2) and target matric potentials (Table 5.3) for irrigation management. Although some of the material is not specific to drip irrigation management, similar alternatives must be considered for drip irrigation scheduling schemes as well.

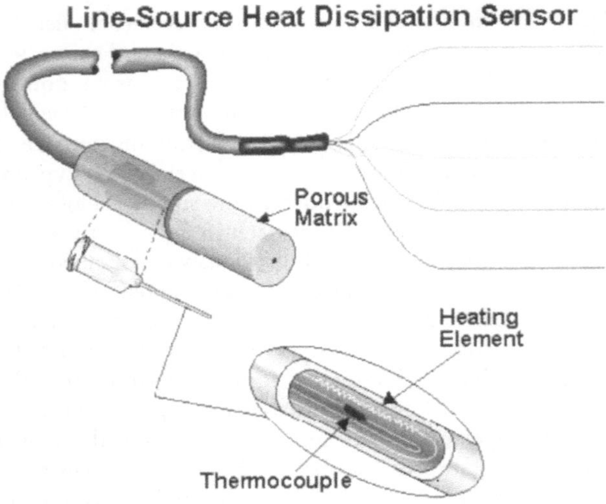

Line-Source Heat Dissipation Sensor

Porous Matrix

Heating Element

Thermocouple

Fig. 5.10. Schematic illustration of line-source heat dissipation sensor (Campbell Scientific Inc, Logan, Utah)

5.3.2
Climatic Monitoring of Evapotranspiration

Climatic data can be integrated to give an estimate of potential evapotranspiration. These data can be augmented by crop coefficients in order to arrive at estimates of actual crop water use. CIMIS (California Irrigation Management Information System) is an example of such a system which has been used successfully for irrigation scheduling (Pitts et al, 1995).

In many cases, the information is provided as a free service, and updated data are accessible by a computer via a modem, or through the World Wide Web (WWW). This important development provides irrigators with updated and highly localized information on a fundamental quantity for irrigation scheduling. Thus, irrigation efficiency may be enhanced since irrgators may improve their ability to calculate crop water use for their area and apply correct amounts of water.

The most widely used method for estimating ET from climatic measurements is the *Penman combination method* considered by many as the standard method. Penman (1948) considered evaporation from saturated surfaces, where the (saturated) vapor pressure at the surface is e_s^*. The crucial step in Penman's analysis is the assumption that:

$$\frac{e_s^* - e_a^*}{T_s - T_a} = \Delta,\tag{5.8}$$

where $\Delta = de^*/dT$ is the slope of the saturation vapor pressure vs. temperature relationship, T_s and T_a are the surface and air temperatures, respectively, and e^* is the saturated vapor pressure at the appropriate temperature. The well-known Penman equation combines energy balance with aerodynamic profile considerations to predict the latent heat flux over well-watered (or saturated) surfaces and is given as:

$$LEt = \frac{\frac{\Delta}{\gamma}(R_N - G) + f(u)\left(e_a^* - e_a\right)}{\frac{\Delta}{\gamma} + 1},$$
(5.9)

where $f(u)$ is the wind function, γ is known as the *psychrometric constant* (about $0.067 \, kPa \, °C^{-1}$ at $20 \, °C$), $\Delta = de^*/dT$ is the slope of the saturation vapor pressure vs. temperature relationships, and e^* is the saturated vapor pressure at the appropriate temperature.

The primary advantage of Penman's equation is that it requires measurements of vapor pressure (relative humidity), temperature, and wind speed at one elevation only (usually at a height of $2 \, m$ above the evaporating surface). The wind function in Penman's equation $f(u)$ is often determined empirically and thus may be site-specific. An approximation was provided by Penman for estimating $f(u)$ as:

$$f(u) = a_w + b_w u_{2m}$$
(5.10)

u_{2m} is the wind speed of $2 \, m$ above the surface in ms^{-1}, $a_w = 1$ and $b_w = 0.864$ (Doorenbos and Pruitt, 1977).

The main use of Penman's equation is for estimation of potential evapotranspiration (ETp) where water is not limiting and the resistance to flow of water vapor is negligible. This concept of potential ET provides a reference estimate which is then adjusted and corrected to account for the actual conditions where water is limiting both evaporation and transpiration. Note that Penman's estimate of ET_P cannot exceed free water evaporation under the same weather conditions.

Empirical Methods. There are various empirical methods for estimating ET, most of which are based on measurement of simple climatological attributes which are considered to be correlated (empirically or semi-empirically) with potential ET. Some of these methods are site- specific or require re-calibration for different geographical areas.

The Jensen and Haise Method (1963) – is based on mean air temperature and solar radiation for estimating daily ET_P (mm/day) according to:

$$ET_p = R_s(0.025 \, T + 0.08),$$
(5.11)

where R_S is the evaporation equivalent of total solar (short-wave) radiation (mm/day), and T is mean air temperature (°C).

Pan Evaporation – pans of various sizes and shapes have been used to measure free water evaporation as an estimate of ET_P. When local advection is unimportant, the agreement between pan evaporation and ET_P is good. Plant water use ranges between 30 to 90% of pan evaporation, depending on the growth stage and the percent cover. Data on conversion factors for various regions in the world and for various crops are available and usually apply to a standard type of pan (class A pan –4 feet in diameter). In many cases, there are simple linear relationships between pan data and potential ET, based on Penman's equation. Phene et al. (1989) found that $ET_{pan} = 1.2ET_{penman}$ (in other words pan evaporation was about $20 \pm 5\%$ higher than ETp based on Penman's equation). Based on these and other results, we conclude that the simplicity and robustness offered by the use of pan data make this method an excellent choice for areas with no climatic measurements or areas with extreme heterogeneity in climatic conditions (Doorenbos and Pruitt 1977).

5.3.3
Plant Monitoring

Plant water status can be estimated by measuring leaf water potential with psychrometers, plant canopy temperature with infrared thermometers, or by stem diameter measurements. All these data can be used as sensors for irrigation control (see the review of Phene 1986); but none of them is economically viable. Sap flow measurements are promising as input for irrigation control (Van Bavel 1995). These measurements provide a direct estimate of the plant water use (transpiration) and several sensors can be integrated to provide input for an irrigation controller.

Plant water status monitoring offers an important source of information for irrigation management. The basic premise behind monitoring plant water status is that the plant provides the ultimate indicator for the effectiveness of either a soil- or a climatic-based scheduling scheme. Both are designed to promote the well-being of the plant; therefore, favorable plant response, as indicated by growth rate, pre-dawn potential, crop stress index and other measures, provides an independent check on their success.

An important indicator of plant water status is its leaf water potential, because cell elongation and plant growth are critically dependent on plant water status, which, in turn, is related to soil water status. This dependency is demonstrated in Fig. 5.11 (Acevedo et al. 1971), which shows leaf elongation rate as a function of soil matric potential and leaf water potential.

This plant sensitivity to leaf (and soil) potential provides a sensitive indicator of desired plant water status for maximum growth rate and yield. The standard method for leaf potential measurement is based on pre-dawn and mid-day leaf water potential using a pressure chamber (Scholander apparatus). Many researchers found a strong correlation between pre-dawn leaf water potential and transpiration (Fig. 5.12; Shouse et al. 1982). Hence, this information may be used as a complementary tool for irrigation scheduling. An extensive review on the pitfalls of the chamber pressure method for the measurement of plant water potential was given by Turner (1987).

Canopy temperature is another indicator for plant water status and the onset of water stress. The concept of the crop water stress index (CWSI), introduced by Jackson (1982) uses measurements of crop temperature measured with a handheld infrared thermometer and air vapor pressure deficit, along with standard baselines (for stressed and non-stressed conditions) to assess the value of CWSI. Although a useful tool for gauging crop water status, it provides an indication for irrigation timing only (Howell et al. 1984). Many additional factors can cause an increase in foliage temperature, such as plant disease, nutrient status or salinity stress. A further development of this concept is measuring the canopy temperature by remote sensing from aircraft or satelite. Large areas can be measured with limited resolution. If these data are combined with ground-based measurements of incoming radiation, reasonable estimates of evapotranspiration can be obtained

Fig. 5.11. Rate of leaf elongation as affected: (a) by soil water potential; and (b) in relation to leaf water potential (Acevedo et al. 1971)

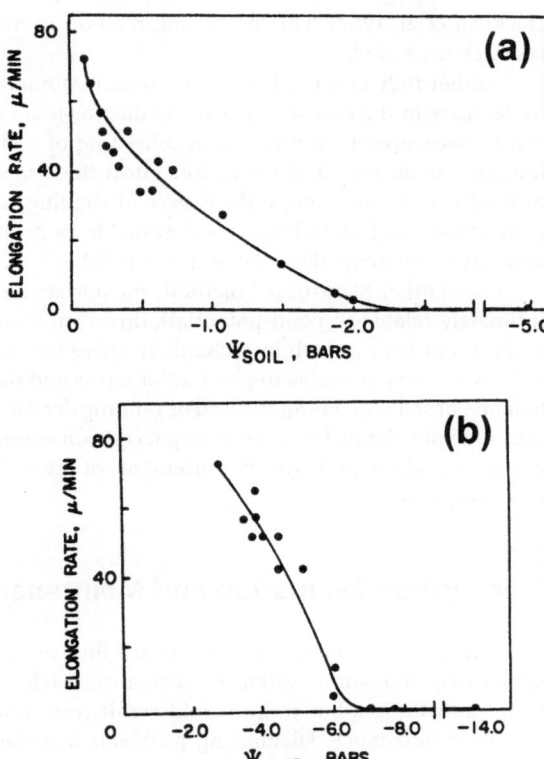

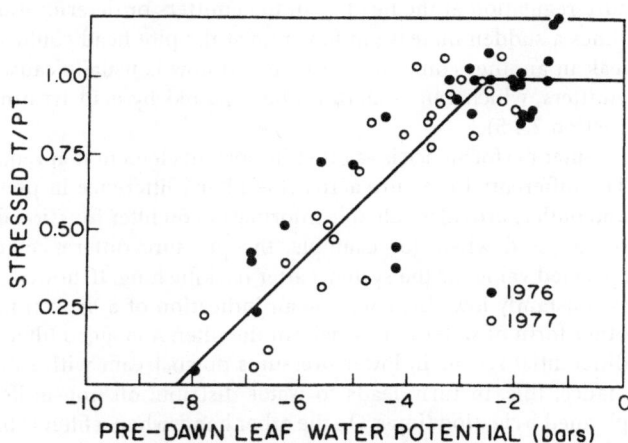

Fig. 5.12. Relative transpiration vs. predawn leaf water potential for cowpeas (Shouse et al. 1982)

(Jackson et al. 1987). This promising method is still in its infancy and much more research is needed.

Another indicator of plant water stress is stomatal resistance. Transpiration tends to decrease in the case of water stress due to stomatal closure. Stomatal conductance can be measured by a porometer, consisting of a chamber attached to a leaf where humidity is measured. Many factors effect these measurements, such as the location of the leaves in the canopy, the degree of shading and the time of day. Sophisticated instruments with data-loggers are available to measure both stomatal conductance and rate of photosynthesis (Parkinson 1985).

Several other plant-based methods include stem diameter measurements (that are intimately related to plant potential), direct measurements of stem elongation rates and rates of fruit growth in orchards (relative to standard, non-stressed rates). These indices are very sensitive to plant water stress and may be useful for checking a water balance-based scheduling method or refining decisions on irrigation timing. However, they are not suitable for determining irrigation amounts (unless there is a way to fully automate and integrate the measurements of several plants and account for the lag in plant response).

5.4
Drip System Evaluation and Maintenance

In section 5.1 we mentioned the possibility of a gap between target design and actual drip irrigation system performance. Such a gap may not necessarily result from poor design, but rather could result from poor management and inadequate system maintenance. Diagnosing problems associated with system performance is usually based on continuous field evaluation using water meters, pressure gauges and spot checks of emitter discharge. The monitoring of flow rates to sub-plots is essential for proper system maintenance and may reveal problems with clogged filters or emitters and permit remedial actions before the problem seriously affects both equipment and yields. A gradual increase in flow rate may indicate inadequate pressure regulation at the head or in the emitters or deterioration of the emitters. Sometimes a sudden increase in flow rate at the plot head could signify a broken line or a leak in another component. Decreased flow is usually caused by clogging of the the emitters, which can sometimes be repaired by acid treatment or chlorination (see Section 2.4.5).

Filter performance is another important element that requires frequent evaluation. The differential pressure across the filter (difference in pressure between filter inlet and outlet) provides valuable information on filter functionality. Filter plugging could be assessed when, for example, the pressure difference remains high (above the specified value for the system) after backflushing. If, however, the pressure difference is constantly low, this could be an indication of a broken filtering element, or some other form of water bypass within the filter. A clogged filter with too high a pressure differential results in lower pressures downstream with sub-optimal emitter performance; this, in turn, leads to water distribution non-uniformity, and longer-than-planned irrigation times. On the other hand, when a filter is broken or non-functional, often the consequences are more severe (i.e., massive emitter clogging). In some situations, however, even with proper filtration design and performance, emitter plugging occurs due to rapid biomass growth (e.g., protozoa). Such cases documented by Ravina et al. (1992) highlight the need for repetitive system evaluation due to the potential for rapid deterioration in system performance despite an apparent optimal design and proper management.

The various components of a drip irrigation system require different preventative and operational maintenance procedures to ensure proper system performance and extend the life of the component. For example, though media filters are considered virtually maintenance-free, a periodic visual inspection of the status of the media could help assess their effectiveness and indicate if the backflushing schedule needs to be adjusted, or perhaps chlorination is needed, due to excess biomass growth. Regularly scheduled maintenance operations, such as the monthly flushing of all laterals, can prevent accumulation of fines and the potential for irreversible emitter plugging. It is advisable to follow the instructions provided by the suppliers of all the components of the drip irrigation system with care. Details on these are given by James and Shannon (1986).

An evaluation procedure for microirrigation system evaluation was developed at Cal Poly (Burt et al, 1985). These evaluations measure the field-wide uniformity for one irrigation event, by measuring pressures and emitter discharge rates. The data are expressed as distibution uniformity [DU, see Eq. (4–17)]. The field performance of 457 microirrigation systems based on this procedure was reported by Hanson et al.(1995). They found distribution uniformities ranging from 30% to 90%, even in brand new systems whose DU was designed to exceed 90%. These evaluations include not only drip systems, but also microsprinklers and sprayers. Only 38% of the systems had a DU above 80%, 35% had a DU below 70% and 28% were between these two values. For the systems with low DU (>70%), clogging was the main contribution to the DU in 18% of the systems, while pressure variation caused low DU in 28% of the systems. The effects of system size, system age and emitter discharge rate (differentiating between drip and others) are shown in Fig. 5-13 from the survey of Hanson et al. (1995). These data are quite disappointing, showing low DU for systems designed for DU above 90%, due to poor management and maintenance.

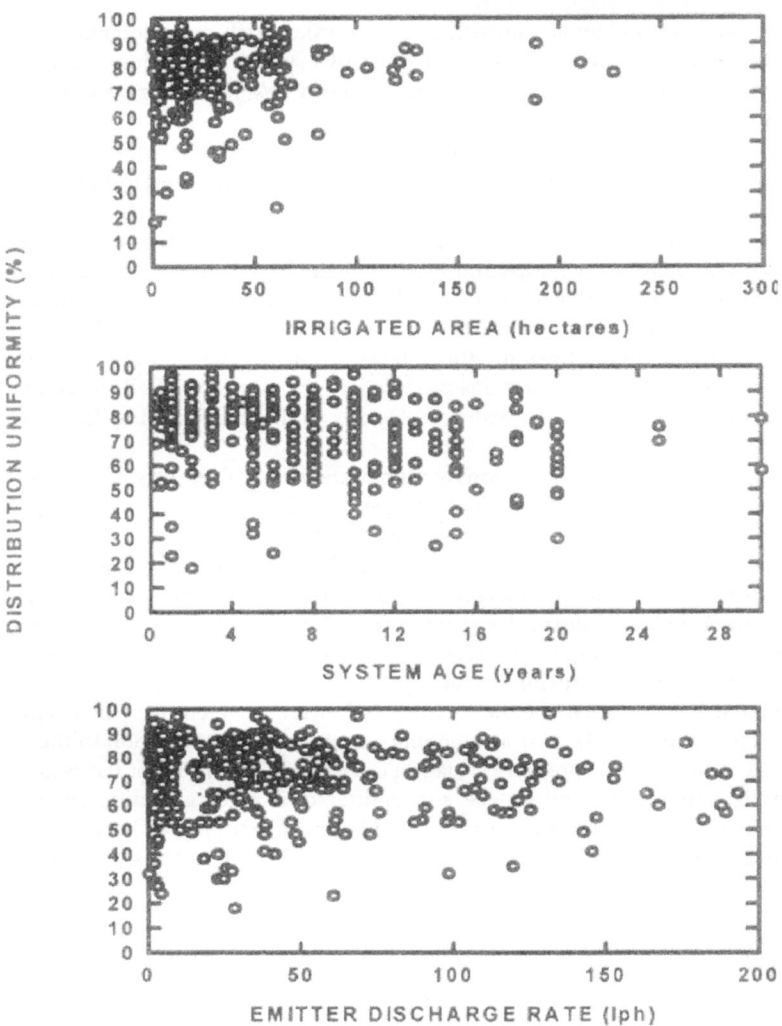

Fig. 5.13. Distribution uniformity (DU) versus irrigated area, system age and emitter discharge rate for 458 microirrigation systems (Hanson et al. 1995)

Practical Applications of Drip Irrigation

In this chapter we will give a critical overview of the applications of drip irrigation to the main branches of agriculture and horticulture. It is our aim to sum up the experience attained by this irrigation method since its evolution and development, and to look at the problems encountered with specific crops under varying circumstances.

6.1
Orchards and Ornamental Trees

Drip irrigation is particularly suitable for tree crops. Trees are usually planted in rows spaced widely enough to allow cultural operations such as spraying, pruning, picking, etc. With traditional irrigation systems (gravity and sprinkler) the whole orchard, including the inter-row spaces, is wetted, thereby interfering with management operations, damaging soil structure, and resulting in soil compaction, particularly if traffic occurs soon after wetting. Moreover, problems of soil aeration due to excess water may occur because the whole soil surface is wetted. These problems do not exist in drip irrigation since here adequate aeration is provided at the fringes of the wetted zone (Bravdo and Proebsting 1993). Moreover, water is applied only in the tree rows, leaving a dry strip between rows. If there is no rain during the irrigation season, very little weed growth takes place in the inter-row spaces. Fertilizers can be applied in a very precise manner through the drip system. Drip irrigation is also well-adapted to provide supplemental irrigation in humid areas because of the permanent installation, the relatively low pressure requirements and the ease of operation. Bucks (1995) estimated, based on available statistics, that more than 10% of the main tree crops worldwide are irrigated by microirrigation, which is a much higher percentage than the fraction of the total irrigated area (see Table 1.1).

6.1.1
Citrus

The response of citrus to irrigation has been investigated, both under Mediterranean-type climates, with dry summers and rainy winters, and under more tropical conditions, with summer rains. Several authors have reported the evapotranspiration of citrus based on water balance measurements. Van Bavel et al. (1967) estimated the cumulative water use of mature orange orchards in Arizona to be 66% of the annual Class A pan evaporation. Kalma and Stanhill (1972) found this to be 54% in the coastal plain of Israel. Irrigation experiments summarized by Shalhevet et al. (1981) gave similar results.

The traditional methods of citrus irrigation are surface (basin and furrow) overhead sprinkler. More recently, below-canopy sprinkler, minisprinkler, microjet and drip systems have been introduced. The question arose as to the extent that partial wetting provided by microirrigation can satisfy the transpirational demand of the trees and reduce evaporation from the soil? Mantell (1977) reviewed the earlier work on partial wetting of the soil surface in Israel. He found that alternate row irrigation provided similar results to those obtained by irrigating every row, using the same amount of water with a doubled irrigation interval. Dasberg (1995) compared several long-term experiments carried out in farmers' groves. The results are shown in Table 6.1. In all these experiments, sprinkling was applied by low-angle sprinklers below the canopy, wetting 90% of the soil surface, spray was applied by one sprayer per tree and drip by 4l/h emitters 75–100 cm apart. The irrigation frequency was adapted to the wetted volume as given by the irrigation method, with the same amount of water applied for all irrigation methods. These data show that in no case did partial wetting of the surface decrease or increase yields on a long-term basis. The response to applied water was similar, which means that no water saving was achieved by partial wetting; the trees used the same amounts of water in all cases.

Experience has shown that the conversion of the irrigation system of established orchards from complete to partial wetting of the soil surface by drip or minisprinklers (30–40% wetted area) has no adverse effects (Bielorai 1982; Bielorai et al. 1985; Bravdo and Proebsting 1993); the citrus root system, even of mature trees, adapts itself quickly to the smaller wetted volume.

In an irrigation experiment with young Valencia oranges in Arizona, the trees irrigated by drip and basin methods showed a larger growth rate during the first 5 years than the trees irrigated by sprinkler and flood (Rodney et al. 1977). Subsequent yields were significantly greater for trees irrigated by the first two, rather than the last two methods. The drip system was operated daily, while the irrigation interval with the other methods was 1 to 2 weeks. In another experiment with young grapefruit trees, conducted for 9 years in Israel (Leitman et al. 1979), no appreciable differences in tree

Table 6.1. Long-term comparison of irrigation methods and quantities on fruit yields (Dasberg, 1995)

Experiment	Years	Applied	Citrus yield (Mg/ha)			
		Water (mm)	Drip, one lateral	Drip, two laterals	Spray	Sprinkler
Grapefruit	4	630	87	91		83
(Yagev, 1977)		800	99	100		94
Grapefruit	7	521	66	68		69
(Bielorai 1977)		707	80			77
		895	82			83
(Bielorai et al. 1985)	7	575			64	60
		765			71	77
		900			82	76
Oranges		610	58		57	53
(Raber et al, 1990)						
Mineola	6	622	53	64	55	
(Dasberg et al. 1994)		504	53	60	57	

development and yield were observed between drip, sprinkler and minisprinkler, using the same irrigation schedule.

The yield response of a mature grapefruit orchard to sprinkler and drip irrigation with one (Drip I) or two (Drip II) laterals per tree row on a sandy loam soil are shown in Fig. 6.1 (Bielorai 1977). He found that yield, fruit quality and tree development were similar in drip (with wetting of only 30% (one lateral) or 40% (two laterals) of the soil surface) and sprinkler irrigation, provided the irrigation is carried out at short intervals (2–3 days). At greater intervals (7 days) drip irrigation with one lateral gave somewhat lower yields. Similar results were obtained by Yagev (1977).

A comparison of drip, microjet, sprinkler and basin irrigation was made on mature Navel orange trees in South Africa. Results showed similar yields for daily drip irrigation compared with microjet and sprinkler with two-week application intervals (Fouche et al. 1979). Results of an experiment with mature Navel oranges on sandy clay loam soil in California showed no difference in fruit production and quality between drip, furrow and sprinkler irrigation (Aljibury 1981). With drip irrigation, the soil water potential was kept at −10 kPa, while in the furrow and sprinkler plots irrigation was applied when the soil water potential at 30 cm depth reached −70 kPa. Nevertheless, the effective amount of water applied by drip irrigation was smaller than by either sprinkler or furrow.

Not all attempts to introduce drip irrigation in mature orange groves have been successful. Cole and Till (1977) reported unsatisfactory results with one drip lateral

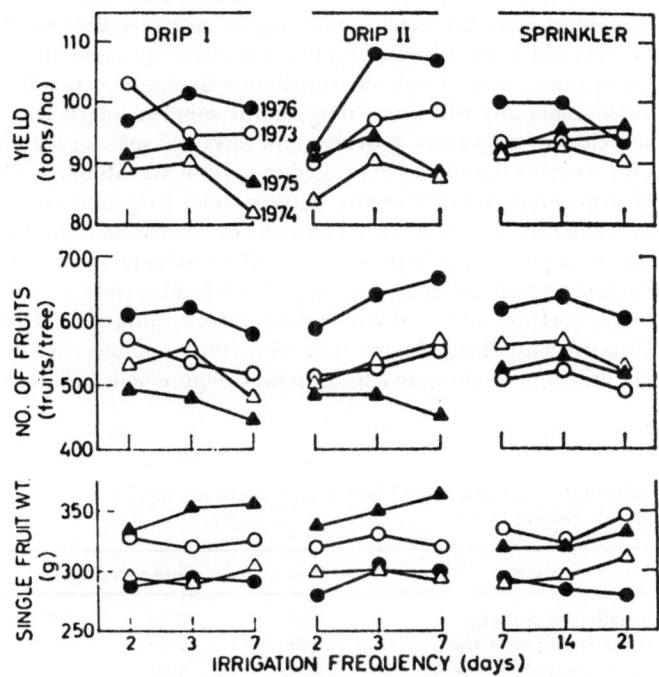

Fig. 6.1. Yield indices of drip and sprinkle irrigated grapefruits at Sa'ad, 1973–1977 (Bielorai 1977)

per row on coarse sand in Australia because only 25% of the area was wetted. The 60% wetting necessary to achieve good production could be attained only by two laterals per row (Cole and Till 1977). In Florida, citrus is planted on sandy soil with low water-holding capacity. Supplementary irrigation is beneficial despite an annual precipitation of 1250–1400 mm. Supplemental irrigation in high-density orange groves (spacing $(2.4 \times 3.4$ m) in Florida on sandy soil was most effective when 70–80% of the area under the tree canopy was irrigated (Koo 1978). In a low-density grove of Valencia oranges, spray irrigation covering 30–50% of the surface area gave higher production than drip irrigation, which covered only 5–10% of the area in a 5-year experiment. Both treatments yielded more than the control without irrigation, although the average precipitation was 1285 mm (Koo and Smajstrla 1985). These studies show the importance of covering sufficient soil area by drip irrigation in a humid area with sandy soil, where irrigation is supplemental. The roots proliferate in the whole soil volume, but drip irrigation wets only part of the root system. In Florida, microsprinklers are often preferred over drip systems, not only because they provide greater soil coverage, but also because they can provide some degree of freeze protection (Boman and Parsons 1995; Smajstrla 1993).

The above results indicate that citrus roots adapt themselves quickly to different patterns of water application, provided the overall transpirational demand of the tree is met. Little information is available on the direct transpiration of citrus trees; therefore, estimating what fraction of the total evapotranspiration results from direct evaporation from the soil is difficult. The available data do not indicate any water saving in orchards by drip irrigation compared with other methods. This implies that, with drip, more frequent water application is necessary than in the traditional methods because of the more limited root volume. Drip appears to be a very good practice for young orange trees. However, with mature trees, adequate soil coverage should be provided, especially when drip irrigation is supplementary. One lateral per tree row is sufficient in most cases, provided flow rates and spacing are adjusted to the hydraulic properties of the soil, as explained in section 4.2. Aljibury (1981) estimated the cost of drip compared with furrow and sprinkler irrigation for citrus in California. The drip and furrow methods used virtually the same small amounts of energy per hectare for direct pumping, whereas the sprinkler system was the highest energy user. A comparison of drip, under-tree spray and overhead sprinkler for supplemental irrigation in Florida (Myers 1977) showed initial and operating costs of a drip system to be lower. Drip irrigation had a higher irrigation efficiency, lower water use and lower operating pressure, resulting in a lower power requirement (see Table 6.2).

Table 6.2. A comparison of irrigation systems for supplemental irrigation of orchard crops in Florida (Myers 1977)

Type of system	Drip	Under tree spray	Overhead sprinkler
Installation costs/ha	865$	1235$	2965$
Water supply costs/ha	250$	450$	531$
Operating pressure (kPa)	83	104	380
Irrigation efficiency (%)	95	85	70
Labor (man h/ha year)	15	5	4

Irrigation generally causes an increase in fruit number and size. Excess irrigation, on the other hand, causes excessive vegetative growth and production of small fruit high in juice and of low keeping quality. Medium water stress (deficit irrigation) does not cause any yield reduction, but tree growth is impaired. Severe stress causes lower fruit set and lower yields (Hilgeman 1977; Castel and Buj 1990; Dasberg and Erner 1995).

6.1.2
Avocado

Avocado culture is developing rapidly in South America, Mexico, Florida, California, South Africa, Israel, North Africa and Spain. One of the first drip irrigation experiments, carried out in an avocado orchard in San Diego County (Gustafson et al. 1980), compared drip with sprinkler irrigation in a commercial orchard on sloping land. They used one sprinkler per tree with water applied weekly, or three emitters (with discharge of 4 l/h) per tree with water applied daily. The water application was regulated by tensiometers at 30 and 60 cm depth. The drip system performed satisfactorily on the sloping land, using appreciably less water in the early years. Tree growth was similar for the two methods. However, fruit yield was somewhat higher for the sprinkler-irrigated trees, probably because the soil volume wetted under drip was too small. Another experiment comparing drip with sprinkling for avocado on a heavy soil was carried out in Israel (Levinson and Adato 1991). They found the dry drip treatment, irrigated intermittently during daylight ($0.46 E_{pan}$), to be superior to the wet drip treatment, irrigated similarly with larger amounts of water ($0.64 E_{pan}$), and to the sprinkler treatments irrigated every 4 days. They related these differences to better root development and better soil aeration (Levinson and Adato 1991). An experiment comparing three levels of irrigation with drip (0.3, 0.6 and $0.9 E_{pan}$) was carried out in Crete. The low irrigation treatment showed lower root density, insufficient wetted volume and lower yields than the other treatments; the high irrigation treatment showed excessive deep percolation and similar yields as the $0.6 E_{pan}$ treatment (Michelakis et al. 1993).

A long-term experiment, on a clay soil in the Western Galilee on trees planted in 1974, was begun in 1978 with four amounts of water applied daily – 60, 80, 100 and 120% replenishment of soil water depletion from the 60-cm soil depth. Tree growth and yields were higher as the water application was increased during the last 5 years of the experiment when the trees had reached full bearing. The main conclusion was that, even with the highly efficient daily drip irrigation, no water saving could be achieved (Lahav and Kalmar 1991).

Another experiment was carried out on trees planted on a sandy loam loessial soil in the northern Negev area. Microjets (one per tree) were compared with drip irrigation (eight emitters with a discharge of 4 l/h per tree) using different amounts of water. The results of the 10-year experiment showed greater tree growth and higher yields with larger amounts of water applied by drip compared with microjet irrigation. Drip irrigation also caused less leaf tip burn and chlorosis than microjet irrigation (Tomer et al. 1995).

6.1.3
Deciduous Trees

Deciduous trees are grown in both temperate and semiarid conditions. An experiment with apples, comparing daily drip irrigation with daily and bi-weekly sprinkler irrigation was carried out at Prosser, Washington, where the summers are dry (Middleton et al. 1979). The amount of water applied to the drip-irrigated trees was determined by the water use of trees grown in lysimeters, while the amount applied by sprinkler was determined by daily pan evaporation. The results showed earlier flowering under drip (in the third and fourth years) than under sprinkler irrigation (see Table 6.3). Consequently, during the first 6 years of growth, yields were much higher under drip than under sprinkler: trees used less water, developed better shapes and required less pruning and spreading. Moreover, drip-irrigated trees yielded fruit of higher quality (both for sale as fresh produce and for preserving) than sprinkler irrigated trees (Drake et al. 1981). This is one of the few well-documented cases showing a specific advantage of drip versus sprinkler under the same irrigation frequency. The authors attributed the differences to changes in root proliferation. The drip-irrigated trees were supplied with one emitter per tree, having a discharge of 4 l/h, which limited the root volume per tree to $0.7 \, m^3$, as compared to a maximum possible root volume of $13 \, m^3$ per tree (Proebsting et al. 1977).

The response of young almond trees to different initial levels of drip irrigation (ranging from 209 to 562 mm) was evaluated in Spain. Trunk growth, canopy area and yield were all linearly related to the amount of irrigation water applied (Torrecilas et al. 1989). Another drip irrigation experiment on young almond trees was carried out in California, with a larger range of water applications (269 to 913 mm annually, during the fifth through seventh year after planting). Trunk growth was a linear function of the total amount of water applied during the experiment. However, yields increased only up to an intermediate level of water application (790 mm, Hutmacher et al. 1994). These data show that, in almonds, with their short growing season, fruit yield is closely related to canopy development.

Drip is suitable for supplemental irrigation in areas with summer rainfall. In dry seasons, drip-irrigated peach trees gave higher yields than sprinkler-irrigated (once, before harvest) trees and non-irrigated controls. The practice of drip irrigation according to evaporation demand also resulted in a greater increase in trunk diameter (Reeder et al. 1979). Similar results were obtained with peach trees in Georgia (Chesness and Couvillon 1980). These authors used four emitters per tree with

Table 6.3. Water use, growth and yields of Redspur Delicious apples under three water regimes (Middleton et al. 1979; Evans and Proebsting 1985)

Irrigation method	Drip, daily	Sprinkler, daily	Sprinkler, bi-weekly
Water application 1973–1976 (cm)	34	304	220
Water application 1977–1978 (cm)	53	159	150
Trunk area increase 1976 (cm²)	14.8	17.7	16.5
Fruit yield 1976 (kg/tree)	8.8	1.0	0.4
Fruit yield 1978 (kg/tree)	45.7	19.5	8.4
Average yield 1978–1984 (kg/tree)	51.3	38.2	–

Q = 4 l/h. They showed the importance of injecting nitrogen into the irrigation water, thus saving appreciable amounts of fertilizer.

Drip installation for supplemental irrigation in cases of drought or deficient rainfall was reported to have definite economic advantages over overhead sprinklers (Funt et al. 1980). Installation costs, pumping costs and water requirements were all less than half of those for sprinklers. Moreover, larger areas can be covered by drip, compared with sprinklers with the same water flow rate. Kipp (1992) reviewed the development of irrigation and fertilization in apple orchards in the Netherlands. Drip irrigation seemed to be an excellent method for supplementary irrigation in dry years. Fertigation resulted in strong shoot growth and larger production because it optimized the nutrient availability in the root zone (Kipp 1992).

6.1.4
Vineyards

Drip irrigation is particularly suitable for grapevines, since the water can be applied to the vine rows without interfering with cultural operations and without wetting the fruit and foliage. An experiment on young vines comparing daily drip with sprinkler and flood irrigation in California (Peacock et al. 1977) showed no difference between the development of the vines as measured by trunk circumferences with all three methods. Water use was much less with drip irrigation, however. During the first year of fruit bearing, no differences occurred in yield and quality between the three methods. Another experiment with older grapevines was carried out in British Columbia (Stevenson 1981). The transition from weekly sprinkler to daily drip irrigation did not cause any changes in growth or yield. It was shown that nitrogen can be applied through the drip system and that the water use efficiency is improved by this irrigation method.

An extensive investigation comparing several drip irrigation modes (three frequencies, one or two emitters per vine and three water quantities) with conventional furrow irrigation with the same three quantities of water was carried out in Arizona (Bucks et al. 1985). Some of the results are given in Table 6.4. These data show that yields and fruit quality were slightly increased with two emitters per vine over furrow irrigation, whereas yield with drip irrigation with one emitter per vine did not differ from furrow irrigation. A trend of increased yield for 3- or 6-day frequencies over daily drip irrigation occurred in 3 out of 4 years. Plant water stress developed and yield was decreased with water applications at 25% less than the seasonal water application (Bucks et al. 1985). It can be concluded that drip irrigation is a suitable method for the growth of vines for grapes as well as for wine production. A large percentage of vines worldwide are drip irrigated.

6.2
Field and Fodder Crops

6.2.1
Cotton

Cotton is a major irrigated crop grown in tropical, sub-tropical and arid regions. Most of the irrigated cotton areas in the world use gravity methods, although in the USA

Table 6.4. Comparison of grape yield for drip and furrow irrigation (Bucks et al. 1985)

Comparison	1973	1974	1975	1976
Two emitters vs. one emitter per vine	+9%	+9%	+17%	NA
Two emitters vs. furrow irrigation	+13%	+7%	+18%	NA
One emitter vs. furrow irrigation	NS	NS	NS	NA
Different drip irrigation quantities	NS	+12% for 1.35 & 1.05 over 0.8 ET	+15% for 1.25 and 1.0 over 0.75 ET	+23% for 1.25 and 1.0 over 1.5 ET
Different furrow irrigation quantities	NS	Trend for 1.35 and 1.05 over 0.8 ET	+29% for 1.25 & 1.0 over 0.75 ET	NS
Different drip frequencies	Trend for 3 and 6 days over daily	Trend for 3 and 6 days over daily	NS	+13% for 6 days over daily

NA = no comparison made, NS = no significant differences, ET = seasonal evapotranspiration

and Israel, sprinkler irrigation is used extensively. In recent years, however, drip has been introduced in Israel and has replaced sprinkling to a large extent, as will be shown below. Wierenga (1977) compared surface and drip irrigation in New Mexico and found that drip irrigation resulted in 8% higher cotton yields than did surface irrigation, using 24% less water at 100% efficiency. Subsequently, drip irrigation was compared with the conventional level-basin method in Arizona (French et al. 1985). It was found that both methods, when properly managed, have a potential for high cotton yields and water use efficiency. Laterals every second row (2 m apart) were satisfactory, but one lateral for every third cotton row was inadequate on these course-textured soils (French et al. 1985). Frequent drip irrigation during fruiting enhanced yields because the period of flowering was lengthened. This could also be achieved by doubling the number of flood irrigations (Radin et al. 1992). Subsurface drip was developed on a commercial scale in Arizona (Tollefson 1985). Drip-tapes were buried at 25 cm depth under each cotton row. This reduced salt concentration near the plant rows, and enabled crop rotation, fertilizer injection and computer control, resulting in appreciable water savings. Drip-irrigated fields out-yielded furrow irrigation by a 30% average over four years (Tollefson 1985). Subsurface drip irrigation of cotton was also developed in Texas, resulting in increased yields over surface irrigation and decreased labor, cultivation and herbicide costs (Hengeller, 1995). An additional benefit of drip irrigation is the possibility of growing cotton by frequent irrigation with highly saline water (Ayars et al. 1985). Good results were also obtained by drip irrigating cotton on saline soils in Spain (Fereres et al. 1985). In the southeastern coastal plain of the USA, cotton irrigation is supplementary. An irrigation system for humid areas should have a low labor requirement, low capital investment and the capacity to sustain the crop during drought at critical periods. Subsurface drip irrigation, using the same laterals for many years and wide lateral spacing, can answer these requirements, as has been shown by several field experiments (Camp et al. 1995b).

Table 6.5. Survey of drip and sprinkle irrigation of cotton in two regions in Israel (Cohen 1978–1986)

Region	Year	Irrigated area (ha) Sprinkle	Drip	Seed cotton yield (kg/ha) Sprinkle	Drip	% Difference
Lachish	1981	4000	500	4540	5120	13
	1982	3075	1100	4820	5320	10
	1983	2900	1750	4990	5270	7
	1984	800	1600	4400	4960	13
Gallilee	1983	7960	1900	5180	5580	8
	1984	3410	2400	4200	5380	28
	1985	2570	3100	4690	5550	18
	1986	1330	3600	4960	5410	9

Israel, although producing only a small fraction of the world cotton crop, produces record yields (see Table 6.5). Cotton is the main irrigated crop in Israel, occupying a large portion of irrigated areas. Until recently, most of the cotton was irrigated by sprinkler, using tow-moved lines, with the water requirements and scheduling determined by extensive experimentation (Shalhevet et al. 1981). During the past decade, drip has replaced sprinkler irrigation on most of the fields. The annual reports of the Cotton Research Board show the development of drip irrigation in cotton (Cohen 1978–1986). Originally, drip was applied mostly on shallow soils with low water-holding capacity, stony soils, coarse-textured soils, fields of irregular size or on steep slopes which were difficult to irrigate with the tractor-towed sprinkler lines. Field experiments showed that drip irrigation resulted in somewhat higher yields than sprinkler with the same amounts of water, usually about 10% more in seed cotton. This comparison is valid only under the most favorable conditions for sprinkler, i.e., irrigation without wind (preferably at night) and with adequate pressure. The drip system for cotton is usually designed with emitters having $Q = 4 l/h$, spaced 1 m apart along the row, one lateral for every two rows; while the sprinklers are generally spaced 18×12 m. No advantages were found in drip irrigation at high frequency (daily or several times daily) compared with low application frequency (once or twice a week). More recent experiments have shown that drip irrigation at the same frequency as sprinkler (every 2–3 weeks) still has some advantages. The increased yield obtained with drip was attributed to several factors:

1. Better water distribution, i.e., less variability in water application
2. More control on precise water and fertilizer application
3. More flexibility in irrigation; larger areas can be irrigated with the same water discharge rate and at a lower pressure compared with sprinkling
4. Irrigation can be carried out at all hours, independent of wind
5. Less labor-intensive, once the system is installed
6. More flexibility in insect and weed control.

Table 6.5 gives the results of surveys carried out in commercial farmers' fields by extension officers, showing a consistent yield increase for drip and a gradual replacement of sprinkling by drip. An economic analysis of cotton irrigation technologies has shown that, under Israeli conditions, drip irrigation is more efficient than sprinkling, linear movement or center pivot, even without taking into consideration the possible yield increase with drip (Rymon and Fishelson 1988).

Drip irrigation of cotton with water of different salinity (3.7 vs 7.8 dS/m) was compared with sprinkler with irrigation amounts of 230–780 mm at different irrigation intervals and for two intra-row spacings (Meiri et al. 1992). It was found that all data fell on a single linear response curve relating seed cotton yield to the sum of applied water and the depletion in soil stored water. With drip irrigation larger amounts of water could be applied without drainage below the root zone, resulting in higher yields (Meiri et al. 1992).

Subsurface drip irrigation was compared to surface drip irrigation for cotton in Israel by Plaut et al. (1985). They found similar yields for both systems when adequate water was supplied. In the case of deficit irrigation, however, subsurface drip was superior since most of the soil evaporation was eliminated (Plaut et al. 1985).

6.2.2
Other Row Crops

Drip irrigation is suitable for row crops, but in many cases the investment for its installation may be relatively high while the economic advantage over other methods may not be apparent. For example, 10 000 m of lateral tubing is required for a field of 1 ha with 1 m row spacing in which a lateral is placed next to each row of plants. In most cases, this high expenditure cannot be compensated by increased crop yields. Therefore, the prevalent drip system in row crops consists of irrigating every second row, with some changes in row spacing if necessary.

Experiments carried out with sorghum in lysimeters showed no significant differences in yield or water use efficiency for drip and sprinkler at high frequency (Ravelo et al. 1977). They also found that variation in drip irrigation frequency (once, twice or three times a week) had no effect on sorghum yield and water use. Bucks et al. (1982) present data showing the same lack of crop response to increased frequency of drip irrigation. Similar results were obtained with corn both in Texas and Kansas. Daily or weekly, sub- or top-irrigated corn gave similar yields if adequate water was applied (Howell et al. 1997; Lamm et al. 1995). High irrigation frequency may result in shallow roots, and a higher crop drought sensitivity. In cases of system breakdown (i.e., mechanical or power failure of the pumping system), the crop irrigated at high frequency will suffer most. A permanent drip irrigation system makes frequent irrigation possible. The frequency should be adapted to the soil (depth and water-holding capacity), the crop (root system development and drought sensitivity), the climate (evaporative demand), and the water quality (salinity). Circumstances under which drip seems to be especially effective include supplemental irrigation in areas with erratic rainfall. Phene and Beale (1976) showed that drip at high frequency gave higher corn yield and less water use than furrow or sprinkler irrigation at the same frequencies. With furrow or sprinkler irrigation, no fine control on water application is possible, and, thus, together with rainfall, excess amounts of water may be applied, resulting in soil aeration problems. Fertilizers should be applied to the irrigation water at an adequate rate. A system of twin-row spacing, developed for these circumstances, requires 40% less irrigation tubing than conventional spacing (Phene and Beale 1979). For other row crops, as with cotton, water saving with drip irrigation compared with other systems is primarily implemented by high water application efficiency and better control.

6.2.3
Sugar Cane

Sugar cane in Hawaii is an example of a crop where drip is taking over as the principal irrigation method. In 1975, less than 10% of the sugar-cane area in Hawaii was irrigated by drip, while in 1979 more than 40% of the total area was drip irrigated. The forecast for 1985 was that 80–90% of all the plantations will convert to drip (Bui and Kinoshota 1985). This rapid conversion has occurred because of cost savings, yield increase and water conservation (Gillespie 1980). Gibson (1976) summarized the early development of drip irrigation of sugar-cane. In 1940, the furrow-irrigated cane fields started to be converted to sprinklers in order to reduce operating costs. In 1970, when above-canopy sprinkling was practiced on about 10% of the cane area, the development of drip irrigation began. The cane crop is harvested 24 months after planting, at which time the drip laterals are discarded. In general, disposable twin-wall tubing is used, with laterals placed between two closely-spaced plant rows (91 cm apart), while the next pair of rows is 183 cm away. The system is operated 12 h daily, applying 8–10 mm water. The clogging problem was solved by adequate filtration and chlorination (Gillespie 1980). Drip irrigated fields have yielded 15–25% more than furrow irrigated plots, and the capital costs for drip are lower than those for sprinkler installations. Lower water application and labor requirements, higher water and nutrient efficiencies and improvements in other operations, such as weed control, are additional reasons for the rapid development of drip for sugar-cane in Hawaii (Bui and Kinoshota 1985). In other sugar-producing areas, the same development is taking place, although at a slower pace.

6.2.4
Forage Crops

Drip irrigation is not practical for most irrigated forage crops because the complete cover of the crop requires relatively close spacing of the laterals, making installation very costly. Furthermore, the harvesting of the forage may damage the irrigation equipment. An exception is the growing of corn as a forage crop on small farms, as practiced in Israel. Corn can give very high yields of dry matter (20–30 Mg/ha) which makes it a very attractive forage crop. The use of overhead sprinklers for corn in small plots is cumbersome, since the height of the sprinklers has to be adjusted as crop height varies. Some of the water is lost at the borders of the plots and wind interferes with irrigation. The drip system for forage corn developed by Leshem (1981) divided the field into 0.5–1 ha subplots, which are planted consecutively at 20-day intervals. The first irrigation is carried out by sprinkler. Then, between 20 and 65 days after planting, the crop is irrigated by drip every 3 to 4 days. The distance between the laterals is 100–200 cm, depending on the soil and planting pattern. Before cutting the crop after 65 days, the laterals are moved to the next plot. Each set of laterals can irrigate three to four plots per season. The amount of water (to which fertilizer is added) for each irrigation is determined by pan-evaporation, the pan factor being adjusted to the growing stage of the crop. The dry matter yield of forage corn obtained with this drip system was similar to that obtained with sprinkler irrigation but with a lower water application (180 mm) than with sprinkler (450 mm) (Leshem 1981).

6.3
Vegetables

Vegetables are high-value crops, generally very sensitive to water stress, especially with respect to yield quality. Moreover, most vegetables are shallow-rooted (Carr, 1981). Therefore, any irrigation system providing a constant supply of water without the development of water stress is suitable for vegetable growing. Drip irrigation is most suitable for vegetable growing since it does not interfere with harvesting operations, fertilizers can be supplied through the system, foliage is not wetted, and water can be supplied frequently and accurately. With frequent drip irrigation, use of brackish water is possible. A relatively low salt concentration is maitained in the root zone, while avoiding foliar damage (Bernstein and Francois 1973). Many experiments with vegetable crops have been reported, of which only a few will be discussed to emphasize a few major points. Lettuce is a very shallow-rooted crop and can give high yields, provided water can be applied frequently. Comparable lettuce yields were obtained with drip, sprinkler, sub-surface and furrow irrigation at high frequency (Sammis 1980). With potatoes, the highest yields were obtained with subsurface irrigation, compared with the other three methods, which gave similar yields (Sammis 1980). Shalhevet et al. (1983) obtained similar potato yields under drip and sprinkler irrigation with the latter requiring 10% more water. They attributed the higher water requirement to evaporation during sprinkling.

One of the first experiments with drip irrigation was carried out in Arizona with cabbage (Bucks et al. 1974). It was found that the consumptive use of water with drip and furrow irrigation at 12-day intervals was the same, and yields were similar. Increasing drip frequency to every 6 or every 3 days gave no increased yield, whereas application of water at less than consumptive use decreased production. With tomatoes, on the other hand, although no differences were observed between drip and furrow irrigation at the same frequencies, yields were higher when water was applied more frequently (Freeman et al. 1976). The effect of irrigation frequency and subsequent development of water stress seems to be specific for each crop and related to soil properties. For beans it depends on the plant growth period; irrigation frequency during the pre-flowering period did not influence pod yield. The highest yield was produced under frequent drip irrigation after flowering. Weekly furrow irrigation during pod set also resulted in a yield increase over bi-weekly irrigation (Muirhead and White 1981). For tomatoes growing on a sandy desert soil, daily drip resulted in much higher yields and growth than did sprinkler irrigation every 3 days. These differences could be attributed to marked lowering of soil and plant water potential (Goldberg et al. 1976). In strawberry culture, the drip laterals can be placed below the plastic mulch, which is usually placed underneath the plants. This practice results in considerable water saving and increased production compared with overhead sprinkling. Locascio et al. (1977) have shown that for best results at least 50% of the fertilizer should be applied with the irrigation water. Some of their results are shown in Table 6.6.

Table 6.6. Effect of irrigation method and timing of fertilizer application on yield of marketable strawberries (Locascio et al. 1977)

Irrigation method	No/ha ($\times 10^3$)	Yield (Mg/ha)	Size (g/fruit)
Drip (daily)	1779	18.85	10.64
Overhead sprinkle (weekly)	1755	16.83	9.77
Control (rainfall only)	1379	13.80	10.00
Fertilizer (%) applied with drip			
0%	1508	16.49	10.91
50%	1902	20.47	10.76
100%	1794	18.42	10.38
Frequency of fertilization			
Daily	1806	19.50	10.78
Weekly	1888	19.39	10.37

6.4
Subsurface Irrigation

In its early development, drip irrigation used laterals with emitters buried in the soil (Blass 1971). This practice was soon discarded because of clogging and root penetration problems. Recently, this practice has been revived for several reasons:

1. The development of more reliable emitters less prone to clogging and new solutions to root penetration problems (treflan)
2. The possibility of using the same system for several consecutive crops, thus saving costs
3. Reducing the labor requirements involved in removing and re-installing the tubing for each crop
4. Less interference with weeding, spraying or harvesting of the crops by buried laterals than by the surface-placed laterals
5. Minimizing water loss by evaporation from the soil surface.

Successful experiments with subsurface drip irrigation were carried out by Bucks et al. (1981). They found no appreciable differences in yields of melon between surface drip, sub-surface drip and furrow irrigation, provided water was applied according to seasonal evapotranspiration demand. Successful crops of onions and carrots were produced with the same subsurface drip system after two previous consecutive crops of melons. No problems of root penetration into the subsurface emitters were encountered. Phene (1995) carried out an experiment comparing high-frequency subsurface drip irrigation (several irrigations per day) with high-frequency surface drip and with low-frequency surface drip on the yield of processing tomatoes. Some of his results are shown in Table 6.7. These data show that injection of a complete fertilizer is essential with high-frequency drip irrigation; yields were almost doubled in 1987, compared with 1984, when only nitrogen was applied. Evapotranspiration (ET_c) was measured continuously by a weighing lysimeter in the SDI treatment and irrigation was applied accordingly. Bare soil evaporation was $0.06ET_c$ for the SDI treatment compared with $0.12ET_c$ for the HFDI treatment. The SDI system has been in operation for more than 10 years. The lack of root intrusion and the maintaining of the original dis-

Table 6.7. Fresh yield of tomatoes and water use efficiency as effected by irrigation method (Phene 1995)

Treatments	1984 (N only) Yield (t/ha)	WUE	1985 (N + P) Yield (t/ha)	WUE	1987 (N + P + K) Yield (t/ha)	WUE
SDI	121a	18a	168a	22a	220a	31a
HFDI	126a	19a	152b	20b	201b	29b
LFDI	114a	16b	130c	18c	187c	26c

WUE = fresh yield/crop evapotranspiration (kg/m^3), SDI = high-frequency subsurface drip irrigation, HFDI = high-frequency surface drip irrigation, LFDI = low-frequency surface drip irrigation

charge of the system was attributed to the high frequency of irrigation, continuous injection of phosphorous acid and yearly injection of fumigant (Phene 1995).

The frequency of SDI had no effect on corn yield in Texas and in Kansas; similar yields were obtained with daily and weekly irrigations (Caldwell et al.1994; Howell et al. 1997; Lamm et al. 1995). A more efficient use of nitrogen can be achieved with SDI rather than by surface drip irrigation (Lamm et al. 1995). Indications of higher P and K uptake by sweetcorn, using subsurface drip fertigation, as compared with surface drip fertigation in a container experiment were reported by Martiez Hernandez et al. (1991). Enhanced root development in the region of high nutrient concentrations was assumed to be the reason for this. Enhanced P uptake with subsurface drip fertigation was not obtained in a field experiment (Bar-Yosef et al. 1989).

Another phenomenon sometimes observed with SDI is a decrease in the discharge of the emitters because of a pressure build-up in the soil (Shani et al., 1996). In soils with low hydraulic conductivity and when the soil is compacted around the emitter this phenomenon may become appreciable, especially with emitters with a high nominal discharge. In light-textured soils and when there is a cavity around the buried emitter, the decrease in emitter discharge is small (Shani et al 1996).

References

Abbottt CL, Koon PDA (1992) Contrasting soil moisture environments beneath sugar cane drip irrigated during the day and at night. Agric Water Manage 22: 271–279

Abramowitz M, Stegun IA (1964) Handbook of mathematical functions. Natl Bur Standards Appl Math Series vol 55. US Government Printing Office Washington, DC

Acevedo E, Hsiao TC, Henderson DW (1971) Immediate and subsequent growth responses of maize leaves to changes in water status. Plant Physiol 48: 631–636

Adin A, Sacks M (1991) Dripper clogging factors in wastewater irrigation. J Irrig Drain Eng ASCE 117: 813–826

Aljibury FK (1981) Water and energy conservation in drip irrigation. In: Annu Techn Conf Proc, Irrigation, the hope and the promise, Irrigation Association. Arlington, Virginia pp 109–115

Aljibury FK, Marsh AW, Huntamer J (1974) Water use with drip irrigation. Proc Second Int Drip Irrig Congr ASAE Publ 105: 341–345

Amoozegar-Fard A, Warrick AW, Lomen DO (1984) Design nomographs for trickle irrigation systems. J Irrig Drain Eng ASCE 110: 107–120

Angelakis AN, Rolston DE, Kadir TN, Scott VN (1993) Soil water distribution under trickle source. J Irrig Drain Eng ASCE 119: 484–500

Anyoji H, Wu IP (1994) Normal distribution water application for drip irrigation. Trans ASAE 37: 159–164

Ayars JE, Hutmacher RB, Schoneman RA, Vail SS, Patton SH, Felleke D (1985) Salt distribution under cotton trickle irrigated with saline water. Proc Third Int Drip/Trickle Irrigation Congress, Fresno, California ASAE 1: 666–672

Barber SA (1984) Soil nutrient bioavailability. Wiley, New York

Bargel C, Baudequin D, Farget H, Penadille Y (1996) Microirrigation dripper performance. Irrigazette 34: 5–8

Bar-Yosef B (1977) Trickle irrigation and fertilization of tomatoes in sand dunes. Water, N and P distributions in the soil and uptake by plants. Agron J 69: 486–489

Bar-Yosef B (1998) Advances in fertigation. In: Sparks DE (ed) Adv Agron 65 (in press)

Bar-Yosef B, Sagiv B, Markovitch T (1989) Sweet corn response to surface and subsurface trickle phosphorus fertigation. Agron J 81: 443–447

Batchelor CH, Soopramanien GC, Bell JP, Nayamuth RA, Hodnett MG (1990) The importance of irrigation regime, dripline placement and row spacing on drip-irrigated sugar-cane. Agric Water Manage 17: 75–95

Batty JC, Hamad SN, Keller J (1975) Energy inputs to irrigation. J Irrig Drain Eng ASCE 101: 293–307

Bell JP, Wellings SR, Hodnett MG, Koon AH (1990) Soil water status in drip-irrigated sugar-cane. Agric Water Manage 17: 171–187

Ben-Asher J, Lomen DO, Warrick AW (1978) Linear and nonlinear models of infiltration from a point source, Soil Sci Soc Am J 42: 3–6

Ben-Asher J, Charach C, Zemel A (1986) Infiltration and water extraction from a trickle irrigation source: the effective hemisphere model. Soil Sci Soc Am J 50: 882–887

Bengson SA (1977) Drip irrigation to revegetate mine wastes in an arid environment. J Range Manage 30(2): 143–147

Bernstein L, Francois LE (1973) Comparison of drip, furrow and sprinkler irrigation. Soil Sci
    115: 73–86
Bernstein L, Francois LE (1975) Effects of frequency of sprinkling with saline waters compared
    with daily drip irrigation. Agron J 67: 185–190
Bielorai H (1977) The effect of drip and sprinkler irrigation on grapefruit yield. Proc Int Soc
    Citric 1: 99–103
Bielorai H (1982) The effect of partial wetting of the root zone on yield and water use efficiency
    in a drip and sprinkler irrigated mature grapefruit grove. Irrig Sci 3: 89–100
Bielorai H, Vaisman I, Feigin A (1980) Drip irrigation of cotton with treated municipal effluents.
    J Environ Qual 13: 231–234
Bielorai H, Dasberg S, Erner Y (1985) Long-term effects of partial wetting in a citrus orchard.
    Proc Third Int Drip/Trickle Irrig Congr Fresno, California ASAE 2: 568–573
Bielorai H, Dasberg S, Erner Y, Brum M (1991) The effect of various soil moisture regimes and
    fertilizer levels on citrus yield response under partial wetting of the root zone. Proc Int Soc
    Citric 2: 585–589
Biggar JW, Nielsen DR (1967) Leaching and miscible displacement. Agronomy 11: 254–271
Black JDF (1976) Trickle irrigation – a review. Hortic Abstr 46: 1–7, 69–74
Black JDF, West DW (1974) Water uptake by an apple tree with various proportions of the root
    system supplied with water. Proc Sec Int Microirrig Congr San Diego, California ASAE Publ
    105: 364–367
Blass S (1971) Drip irrigation. In: Drip (trickle) and automated irrigation in Israel. Water Com-
    missioners Office, Ministry of Agriculture, Tel Aviv, Israel 1: 10–28
Boman B, Parsons ML (1995) Considerations for component selection in microsprinkler selec-
    tion. Proc Fifth Int Microirr Congr Orlando, Florida ASAE Publ 4: 701–706
Bos MG (1979) Standards for irrigation efficiencies of ICID. J Irrig Drain Eng, ASCE 105(IRI):
    37–43
Bouyoucous GJ, Mick AH (1940) An electrical resistance method for continuous measurement
    of soil moisture under field conditions. Mich Agric Exp Stn Tech Bull 172
Bralts VF (1983) Hydraulic design and field evaluation of drip irrigation submain units, Ph D
    thesis, Department of Agricultural Engineering, Michigan State University, Ann Arbor
Bralts VF, Wu IP, Gitlin HM (1981) Manufacturing variation and drip irrigation uniformity. Trans
    ASAE 24: 113–119
Bralts VF, Gerrish PJ, Yue R (1995) An improved finite model for the analysis of microirrigation
    systems. Proc Fifth Int Microirrigation Cong Orlando, Florida ASAE Publ 4: 651–655
Brandt A, Bresler E, Diner N, Ben Asher J, Heller J, Goldberg D (1971) Infiltration from a trickle
    source. I. Mathematical models. Soil Sci Soc Am Proc 35: 675–683
Bravdo B, Proebsting EL (1993) Use of drip irrigation in orchards. Hortic Technol 3: 44–50
Bresler E (1975) Two-dimensional transport of solutes during nonsteady infiltration from a
    trickle source, Soil Sci Soc Am Proc 39: 604–613
Bresler E (1977) Trickle-drip irrigation: Principles and application to soil water management.
    Adv Agron 29: 343–393
Bresler E (1978) Analysis of trickle irrigation with application to design problems. Irrig Sci 1:
    3–17
Bresler E, Green RE (1987) Transport of a degradable substance and its metabolites under drip
    irrigation. Agric Water Manage 12: 195–206
Bresler E, Russo D (1975) Two-dimensional solutes transfer during nonsteady infiltration.
    Laboratory test of mathematical model. Soil Sci Soc Am Proc 39: 585–587
Bresler E, Heller J, Diner N, Ben Asher J, Brandt A, Goldberg, D (1971) Infiltration from a
    trickle source: II. Experimental data and theoretical prediction. Soil Sci Soc Am Proc 35:
    683–689
Bresler E, McNeal BL, Carter DL (1982) Saline and sodic soils. Springer, Berlin Heidelberg New
    York
Bristow KL, Kluitenberg GJ, Horton R (1994) Measurement of soil thermal properties with a
    dual probe heat-pulse technique. Soil Sci Soc Am J 58: 1288–1294

Bucks DA (1995) Historical development in microirrigation. Proc. Fifth Int Microirrig Congr Orlando, Florida, ASAE Publ 4: 1–6

Bucks DA, Davis S (1986) Historical development. In: Nakayama FS, Bucks DA (eds) Trickle irrigation for crop production. Elsevier, Amsterdam, pp 1–26

Bucks DA, Erie LJ, French OF (1974) Quantity and frequency of trickle and furrow irrigation for efficient cabbage production. Agron J 66: 53–57

Bucks DA, Erie LJ, French OF, Nakayama FS, Pew WD (1981) Subsurface trickle irrigation management with multiple cropping. Trans ASAE 24: 1482–1489

Bucks DA, Nakayama FS, Warrick AW (1982) Principles, practices and potentialities of trickle (drip) irrigation. Adv Irrig 1: 219–297

Bucks DA, French OF, Nakayama FS, Fangmeier DD (1985) Trickle irrigation management for grape production. Proc Third Int Drip/Trickle Irr Congr Fresno, California ASAE 1: 204–211

Bui W, Kinoshota CM (1985) Has drip irrigation in Hawaii lived up to its expectations? Proc Third Int Drip/Trickle Irrig Congr Fresno, California ASAE 1: 84–90

Burt C, O'Connor K, Ruehr T (1995) Fertigation. Irrigation Training and Research Center, California Polytechnical State University, San Luis Obispo, CA

Burt CM, Walker PE, Styles SW (1985) Evaluation of microirrigation systems. Proc Third Int Drip/trickle Irrig Congr Fresno, California ASAE 1: 268–273

Caldwell DS, Spurgeon WE, Mangess HL (1994) Frequency of irrigation for subsurface drip-irrigated corn. Trans ASAE 37: 1099–1103

Camp CR, Sadler EJ, Busscher WJ, Sojka RE, Karlen DL (1995a) Experiences with microirrigation for agronomic crops in the southeastern USA. Proc Fifth Int Microirrigation Congr Olando, Florida ASAE Publ 4: 638–644

Camp CR, Hunt PG, Bauer PJ (1995b) Subsurface microirrigation management and lateral spacing for cotton in the southeastern USA. Proc Fifth Int Microirrig Congr Orlando, Florida ASAE Publ 4: 368–374

Campbell GS, Campbell MD (1982) Irigation scheduling using soil moisture measurements: theory and practice. Adv Irrig 1: 25–42

Carr MKV (1981) The role of irrigation in vegetable production. In: Spedding CRW (ed) Vegetables for feeding people and livestock. Macmillan, London

Cary JW, Fisher HD (1983) Irrigation decisions simplified with electronics and soil water sensors. Soil Sci Soc Am J 47: 1219–1223

Castel JR, Buj A (1990) Response of salustiana oranges to high frequency deficit irrigation. Irrig Sci 11: 121–127

Chapin RD (1971) Drip irrigation in the United States. In: Drip (trickle) and automated irrigation in Israel, Water Commissioners Office, Ministry of Agriculture, Tel Aviv, pp 207–216

Chase RG (1985) Phosphorus application through a subsurface trickle irrigation system. Proc Third Int Drip/Trickle Irrig Congr Fresno, California ASAE 1: 393–400

Chesness JL, Couvillon GA (1980) Peach tree response to trickle application of water and nutrients. ASAE paper no 80-2079. American Society of Agricultural Engineers, St Joseph Michigan

Christiansen JE (1942) Irrigation by sprinkling. Univ Calif Agric Exp Stn Bull 670, 124 pp

Clothier BE (1984) Solute travel times during trickle irrigation. Water Resour Res 20: 1848–1852

Clothier BE, Sauer TJ (1988) Nitrogen transport during drip fertigation with urea. Soil Sci Soc Am J 52: 345–349

Cogels OG (1983) An irrigation uniformity function relating the effective uniformity of water application to the scale of influence of the plant root zone. Irrig Sci 4: 289–299

Coelho EF, Or D (1996) A parametric model for two-dimensional water uptake by corn roots under drip irrigation. Soil Sci Soc Am J 60: 1039–1049

Coelho EF, Or D (1997) Applicability of analytical solutions for flow from point sources to drip irrigation management. Soil Sci Soc Am J 61: 1331–1341

Cohen A (1978–1986) Research in cotton growth. Annual Reports, Ministry of Agriculture, Extension Service and Cotton Production and Marketing Council, Tel Aviv (in Hebrew)

Cole PJ, Till MR (1977) Evaluation of alternatives to overhead sprinklers for citrus irrigation. Proc Int Soc Citric 1: 103–106

Curtis AA, Watson KK (1979) Infiltration and redistribution of water from a trickle source. Irrigation Efficiency Seminar, Australian National Committee, ICID, Sydney

Dasberg S (1995) Drip and spray irrigation of citrus orchards in Israel. Proc Fifth Int Microirrigation Congr Orlando, Florida ASAE Publ 4: 281–287

Dasberg S, Bresler E (1985) Drip irrigation manual. International Irrigation Information Center (IIIC) Bet Dagan, Israel Publ 9

Dasberg S, Erner Y (1996) The effects of irrigation management and nitrogen application on yield and quality of Mineola mandarins. Acta Hort 449: 125–136

Davis S, Bucks DA (1983) Drip Irrigation. In: Pair CH, Hinz DD, Sneed RA, Frost KR, Schiltz TJ (eds) Irrigation, fifth edition. Irrigation Association, Silver Springs, Maryland, pp 528–546

Decroix M, Malaval A (1985) Laboratory evaluation of trickle irrigation equipment for field system design Proc Third Int Drip/Trickle Irrig Congr Fresno, California ASAE 1: 325–333

Dhuyvetter KC, Lamm FR, Rogers DH (1995) Subsurface drip irrigation for field corn – an economic analysis. Proc Fifth Intern Microirr Congr Orlando, Florida ASAE Publ 4: 395–401

Doorenbos J, Kassam AH (1979) Yield response to water. FAO irrigation and drainage paper 33 FAO Rome

Doorenbos J, Pruitt WO (1977) Crop water requirements. FAO Irrigation and Drainage Paper 24 FAO Rome

Drake SR, Proebsting EL, Mahan MO, Thompson JB (1981) Influence of trickle and sprinkle irrigation on 'Golden Delicious' apple quality. J Am Soc Hortic Sci 106(3): 255–258

Ellsworth TR, Butters GL (1993) Three-dimensional analytical solutions to the advection-dispersion equations in arbitrary cartesian coordinates. Water Resour Res 29: 3215–3225

Evans RG, Proebsting EL (1985) Response of red delicious apples to trickle irrigation. Proc Third Int Drip/Trickle Congr Fresno, California ASAE 1: 231–239

FAO (1982) Production yearbook (FAO, Rome)

FAO (1992) Production yearbook (FAO, Rome)

Feddes RA, Bresler E, Neuman SP (1974) Field test of a modified numerical model for water uptake by root systems. Water Resour Res 10(6): 1199–1205

Feigin A, Ravina I, Shalhevet J (1991) Irrigation with treated sewage effluent. Springer, Berlin Heidelberg New York

Fereres E, Cuevas R, Orgaz F (1985) Drip irrigation of cotton in southern Spain. Proc Third Int Drip/trickle Irrig Congr Fresno, California ASAE 1: 187–193

Ford HW (1977) The Importance of water quality in drip/trickle irrigation systems. Proc Int Soc Citric 1: 84–87

Ford HW, Tucker DPH (1974) Water quality measurements for drip irrigation systems. Proc Fla State Hortic Soc 87: 58–60

Fouche PS, Bester DH, Veldman GH (1979) A comparison of different methods to increase productivity of Navel oranges. Citrus Subtrop Fruit J 8: 4–6

Freeman BM, Blackwell J, Garzoli KV (1976) Irrigation frequency and total water application with trickle and furrow systems, Agric Water Manage 1: 21–31

French OF, Bucks DA, Roth RL, Gardner BR (1985) Trickle and level-base irrigation management for cotton production. Proc Third Int Drip/Trickle Irr Congr Fresno, California ASAE II: 555–562

Frith GLT, Nichols DG (1974) Effects of nitrogen fertilizer application to part of a root system. Proc 2nd Int Drip Irrig Congr ASAE Publ 105: 212–214

Funt RC, Ross DS, Brodie HL (1980) Economic comparison of trickle and sprinkler Irrigation of six fruit crops in Maryland 1978. Md Agric Exp Stn Rep No 950, 16 p

Gardner WH (1986) Water content. In: Klute A (ed) Methods of soil analysis, vol 1: ASA, Madison, Wisconsin, pp 493–544

Gardner WR (1958) Some steady-state solutions of the unsaturated moisture flow equation with application to evaporation from water table. Soil Sci 85: 228–232

Gerstl Z, Saltzman S, Kliger L, Yaron B (1981) Distribution of herbicides in soil in a simulated drip irrigation system. Irrig Sci 2: 155–166

Gibson W (1976) Hydraulics, mechanics and economics of subsurface and drip irrigation of Hawaiian sugarcane. Int Sugar J 78: 40–44

Gilbert RC, Nakayama FS, Bucks DA (1979) Trickle irrigation: prevention of clogging. Trans ASAE 22(3): 514–519

Gilbert RC, Nakayama FS, Bucks DA, French OF, Adamson KC (1981) Trickle irrigation: Emitter clogging and flow problems. Agric Water Manage 3: 159–178

Gillespie B (1980) Drip irrigation as applied to sugarcane in Hawaii. In: Annu Techn Conf Proc, Irrigation association, Arlington, Virginia, pp 105–112

Goldberg D, Uzrad M (1976) Strip cultivation of the area wetted by drip irrigation in the Arava desert. Hortscience 11(2): 136–138

Goldberg D, Gornat B, Rimon D (1976) Drip irrigation-principles, design and agricultural practices. Drip Irrigation Scientific Publications, Kfar Shmaryahu, Israel

Green SR, Clothier BE (1995) Root water uptake by kiwifruit vines following wetting of root zone. Plant Soil 173: 317–328

Grimes DW, Sohweers VH, Wiley PL (1976) Drip and furrow irrigation of fresh market tomatoes on a slowly permeable soil. II. Water Relations. Calif Agric 76: 11–13

Grobbelaar HL, Lourens F (1974) Fertilizer applications with drip irrigation. Proc 2nd Int Drip Irrig Congr ASAE Publ 105: 411–415

Gustafson CD (1977) Drip irrigation '76 Irrig J 1977: 37

Gustafson CD, Marsh AW, Branson RL, Davis S (1980) Drip irrigation on avocados. Avocado Yearb 1980: 95–135

Hagan RM, Haise HR, Edminster TW eds (1967) Irrigation of agricultural lands. Agronomy 11 ASA, Madison, Wisconsin

Haise HR, Hagan RM (1967) Soil plant and evaporative measurements as criteria for scheduling irrigation. Agronomy 11: 557–604

Halevy I, Boaz M, Zohar Y, Shani M, Dan H (1973) Trickle irrigation. In: FAO Irrigation and Drainage Paper 14: 74–121, Rome

Hanks JR (1983) Yield and water use relationships: an overview. In: Taylor HM, Jordan WR, Sinclair TR (eds) Limitations of efficient water use in crop production. ASA Madison, Wisconsin pp 393–411

Hanson BR (1995) Drip irrigation of row crops: an overview Proc Fifth Int Microir Congr Orlando, Florida ASAE Publ 4: 651–656

Hanson BR, Bowers W, Davidoff B, Kaspligil D, Carvajal A, Bendixen W (1995) Field performance of microirrigation systems. Proc Fifth Int Microir Congr Orlando, Florida ASAE Publ 4: 769–774

Hart WE, Reynolds WN (1965) Analytical design of sprinkler systems. Trans ASAE 8: 83–89

Healy RW, Warrick AW (1988) A generalized solution to infiltration from a surface point source. Soil Sci Soc Am J 52: 1245–1251

Heller J, Bresler E (1973) Trickle irrigation. In: Yaron D, Danfors E, Vaadia Y (eds) Arid zone irrigation. Springer Berlin Heidelberg New York, pp 339–352

Hendrickx JMH, Wierenga PJ (1990) Variability of soil water tension in a trickle irrigated Chile pepper field. Irrig Sci 11: 23–30

Hengeller JC (1995) A history of drip irrigated cotton in Texas Proc Fifth Int Microir Congr Orlando, Florida ASAE Publ 4: 669–674

Hilgeman RH (1977) Response of citrus trees to water stress in Arizona. Proc Int Soc Citric 1: 10–74

Hilhorst MA, Dirksen C (1994) Dielectric water content sensors. Time domain vs. frequency domain. Symposium and workshop on time domain reflectometry in environmental infrastructure and mining applications. Special Publication Sp19-94 Department of Interior, Bureau of Mines, Evanston, Illinois, pp 23–33

Hiller EA, Howell TA (1983) Irrigation options to avoid critical stress; an overview. In: Taylor HM, Jordan WR, Sinclair TR (eds). Limitations to efficient water use in crop production. ASA, Madison, Wisconsin pp 479–505

Hills DJ, El-Ababy FG (1990) Evaluation of microirrigation self-cleaning emitters. Appl Eng Agric 6: 441–445

Hills DJ, Tajirshy M (1995) Treatments requirements of secondary effluent for microirrigation. Proc Fifth Intern Microir Congr Orlando, Florida ASAE Publ 4: 887–890

Hills DJ, Nawar FM, Waller PM (1989) Effects of chemical clogging on driptape irrigation uniformity Trans ASAE 32: 1202–1206

Hodnett MG, Bell JP, Koon AH (1990) The control of drip irrigation of sugar-cane using index tensiometers. Agric Water Manage 17: 189–207

Hoffman GJ, Ayers RS, Doering EJ, McNeal BL (1980) Salinity in irrigated agriculture. In: Jensen ME (ed) Design and operation of farm irrigation systems. ASAE, St Joseph, Michigan

Howell TA, Barinas A (1980) Pressure losses across trickle irrigation fittings and emitters. Trans ASAE 23(4): 928–933

Howell TA, Hiller EA (1974) Trickle irrigation lateral design. Trans ASAE 17: 902–908

Howell TA, Hatfield JH, Yamada H, Davis KR (1984) Evaluation of cotton canopy temperature to detect crop water stress. Trans ASAE 27: 84–88

Howell TA, Schneider AD, Evett SR (1997) Subsurface and surface microirrigation of corn-southern high plains. Trans ASAE 40: 635–641

Hutmacher RB, Nightingale HI, Rolston DE, Biggar JW, Dale F, Vail SS, Peters D (1994) Growth and yield response of almond to trickle irrigation Irrig Sci 11: 117–127

Jackson RD (1982) Canopy temperature and crop water stress. Adv Irrig 1: 43–85

Jackson RD, Moran MS, Gay LW, Raymond LH (1987) Evaluating evaporation from field crops using airborne radiometry and ground-based meteorological data. Irrig Sci 8: 81–90

James LG, Shannon WM (1986) Flow measurement and system maintenance. In: Nakayama FS, Bucks DA (eds) Trickle irrigation for crop production. Elsevier, Amsterdam, pp 280–316

Jarvis NJ (1989) A simple empirical model of root water uptake. J Hydrol 107: 57–72

Jensen ME (1973) Consumptive use of water and irrigation water requirements. ASCI, New York

Jensen ME, Haise HR (1963) Estimating evapotranspiration from solar radiation Irrig Drain Div ASCE 89: 15–41

Jungk AO (1996) Dynamics of nutrient movement at the soil-root interface. In: Waisel Y, Eshel A, Kafkafi U (eds) Plant roots the hidden half. Marcel Dekker, New York, pp 529–556

Jury WA, Earl KD (1977) Water movement in bare and cropped soil under isolated trickle emitters. Soil Sci Soc Am J 41: 852–856

Jury WA, Gardner WR, Gardner WH (1991) Soil Physics. John Wiley and sons, New York NY

Kalma JD (1970) Some aspects of the water balance of an irrigated orange plantation. PhD Thesis, Faculty of Agriculture, Hebrew University, Rehovot

Kalma JD, Stanhill G (1972) The climate of an orange orchard. Physical characteristics and microclimate relationships. Agric Meteorol 10: 185–201

Karmeli D (1977) Classification and flow regime analysis of drippers. J Agric Eng Res 22(2): 165–173

Karmeli D, Keller J (1975) Trickle irrigation design Rain Bird Sprinkling Manufacturing Corporation, Glendora, California, 132 pp

Katzenelson E, Teltch B (1976) Dispersion of enteric bacteria in the air as a result of sewage spray irrigation and treatment processes. J Water Pollut Control Fed 48: 710–716

Keller J, Bliesner RD (1990) Sprinkle and trickle irrigation. Chapman and Hall, New York

Kipp JA (1992) Thirty years fertilization and irrigation in dutch apple orchards. Fertil Res 32: 149–156

Koo RCJ (1978) Response of densely planted Hamlin orange on two rootstocks to low volume irrigation. Proc Fla State Hortic Soc 91: 8–10

Koo RCJ, Smajstrla AG (1985) Trickle irrigation of citrus on sandy soils in a humid region. Proc Third Intern Drip/Trickle Irrig Congr Fresno, California ASAE 1: 212–220

Kutilek M, Nielsen DR (1994) Soil hydrology. Catena, Cremlingen-Destedt

Lafolie F, Guennelon R, van Genuchten MTh (1989) Analysis of water flow under trickle irrigation: I. Theory and numerical solution. Soil Sci Soc Am J 53: 1310-1318

Lahav E, Kalmar D (1991) The effect of cutting the water amounts in a drip-irrigated avocado orchard. Final Report 307/02. Agricultural Research Organization, Bet Dagan (in Hebrew)

Lamm FR, Spurgeon WE, Rogers DH, Manges HL (1995) Corn production using subsurface drip irrigation. Proc Fifth Intern Microir Congr Orlando, Florida ASAE Publ 4: 388-394

Landers JN, Witte K (1967) Irrigation for frost protection. Agronomy 11: 1037-1055

Landsberg JJ, McMurrie R (1984) Water use by isolated trees. Agric Water Manage 8: 223-242

Leij FJ, Dane JH, van Genuchten MT (1991a) A mathematical analysis of one-dimensional solute transport in a layered soil profile. Soil Sci Am J 55: 944-953

Leij FJ, Skaggs TH, van Genuchten MT (1991b) Analytical solutions for solute transport in three-dimensional semi-infinite porous media. Water Resour Res 27: 2719-2733

Leij FJ, Alves WJ, Van Genuchten MT, Williams JR (1999) The UNSDOA unsaturated soil hydraulic conductivity data base. Proc Int Workshop on indirect methods for estimating hydraulic properties of unsaturated soils. US Salinity Lab, Riverside, California

Leitman G, Shufman E, Shugrun A, Shain M, Ashkenazi S (1979) Solid-set irrigation in the citrus grove from the day of planting. Comparing sprinkling, microjets and trickle systems. Hassadeh 60: 767-771 (in Hebrew)

Leshem J (1981) Trickle irrigation of forage corn. Hassadeh 60: 1371-1379 (in Hebrew)

Letey J, Dinar A, Woodring C, Oster JD (1990) An economic analysis of irrigation systems. Irrig Sci 11: 37-43

Levin I, Van Rooyen PC, Van Rooyen FC (1979) The effect of discharge rate and intermittent water application by point source irrigation on the soil moisture distribution pattern. Soil Sci Soc Am J 43: 8-16

Levin I, Sarig S, Meron M (1985) Tensiometer location in controlled automated drip irrigation of cotton. Proc Third Int Drip/tickle Irrig Congr Fresno, California ASAE 2: 782-785

Levinson B, Adato I (1990) Influence of reduced rates of water application using daily intermittent drip irrigation on the water requirements, root development and response of avocado trees. J Hortic Sci 66: 200-216

Locascio SJ, Myers JM, Martin FG (1977) Frequency and role of fertilization with trickle irrigation for strawberries. J Am Soc Hortic Sci 102(4): 456-458

Lomen DO, Warrick AW (1974) Time-dependent linearized infiltration: II. Line sources. Soil Sci Soc Am Proc 38: 568-572

Lomen DO, Warrick AW (1976) Solution of the one-dimensional linear moisture flow equation with implicit water extraction functions. Soil Sci Soc Am J 40: 342-344

Lomen DO, Warrick AW (1978) Linearized moisture flow with loss at the soil surface. Soil Sci Soc Am J 42: 396-400

Lupin M, Magen H, Gambash Z (1996) Preparation of solid fertilizer based solution fertilizers under "grass roots" field conditions. Fertil News FAI 41(12): 69-72

Mantell A (1977) Water application efficiency of citrus: room for improvement? Proc Int Soc Citric 1: 74-79

Mantell A, Frenkel H, Meiri A (1985) Drip irrigation of cotton with saline-sodic water. Irrig Sci 6: 95-106

Marsh AW (1977) Drip irrigation in California. Calif Agric 78: 19

Martinez Hernandez JJ, Bar-Yosef B, Kafkafi U (1991) Effect of surface and subsurface drip fertigation on sweet corn rooting, uptake, dry matter production and yield. Irrig Sci 12: 153-159

McElhoe BA, Hilton HW (1974) Chemical treatment of drip irrigation water. Proc Sec Int Drip Irrig Congr ASAE Publ 105: 215-220

Meiri A, Frenkel H, Mantell A (1992) Cotton response to water and salinity under sprinkler and drip irrigation. Agron J 84: 44-50

Meron M, Asaf R, Bravdo B, Wallach R, Hallel R, Levin A, Dahan I (1995) Soil sensor actuated microirrigation of apple trees. Proc Fifth Int Microirrigation Congr Orlando, Florida ASAE 4–95: 486–491

Merriam JL, Keller J (1978) Farm irrigation system evaluation, a guide for management. Utah State University Logan

Michelakis N, Vougioucalou E, Clapaki G (1993) Water use, wetted soil volume, root distribution and yield of avocado under drip irrigation. Agric Water Manage 24: 119–131

Middleton JE, Proebsting EL, Roberts S (1979) Apple orchard irrigation by trickle and sprinkler. Trans ASAE 22(3): 582–584

Miller RJ, Rolston DE, Rauschkolb RS, Wolfe DW (1975) Drip application of nitrogen is efficient. Calif Agric 76: 16–18

Miller RJ, Rolston DE, Rauschkolb RS, Wolfe DW (1981) Labeled nitrogen uptake by drip-irrigated tomatoes. Agron J 73: 265–270

Mmolawa K (1999) Solute dynamics in drip irrigated fields. Thesis, Utah State University, Logan, Utah

Molz FJ (1981) Interaction of water uptake and root distribution. Agron J 56: 35–41

Muirhead WA, White RJC (1981) The influence of soil water potential on the flowering pattern, pod set and yields of snap beans (Phaseolus vulgaris L.) Irrig Sci 3: 45–56

Myers JM (1977) Functional performance of irrigation system for orchard crops in Florida. Proc Fla Sta Hortic Soc 90: 258–260

Nakayama FS (1986) Water treatment. In: Nakayama FS, Bucks DA (eds) Trickle irrigation for crop production, Elsevier, Amsterdam, pp 164–188

Nakayama FS, Bucks D (1986) Trickle irrigation for crop production. Design, operation and management. Elsevier, Amsterdam

Nakayama FS, Bucks DA, French OF (1977) Reclaiming partially clogged trickle emitters. Trans ASAE 20(2): 279–280

Neumann SP, Feddes RE, Bresler E (1975) Finite element analysis of two-dimensional flow in soils considering water uptake by roots. Soil Sci Soc Am Proc 39: 225–230

Nightingale HI, Phene CJ, Patton SH (1985) Trickle irrigation effects on soil chemical properties. Proc Third Int Drip/Trickle Irrigation Congress, Fresno, California ASAE 2: 730–735

Novak V (1994) Water uptake of maize roots under conditions of non-limiting soil water content. Soil Technol 7(1): 37–45

Or D (1995a) Soil water sensor placement and interpretation for drip irrigation management in heterogeneous soils. Proc Fifth Int Microirrig Congr Orlando, Florida ASAE Publ 4: 214–221

Or D (1995b) Stochastic analysis of soil water monitoring for drip irrigation management in heterogeneous soils. Soil Sci Soc Am J 59: 122–1233

Or D (1996) Drip irrigation in heterogeneous soil. Steady-state field experiments for stochastic model evaluation. Soil Sci Soc Am J 60: 1339–1349

Or D, Coelho EF (1966) Soil water dynamics under drip irrigation: transient flow and uptake models. Trans ASAE 39: 2017–2025

Or D, Groeneveld DP (1994) Stochastic estimation of plant available soil water in Owens valley under fluctuating water depths. J Hydrol 163: 63–64

Or D, Hanks RJ (1993) Irrigation scheduling considering soil variability and climatic uncertainty: Simulation and field studies. In: Russo D, Dagan G (eds) Water flow and solute transport in soils, development and applications. Springer, Berlin Heidelberg New York, pp 262–282

Oron G, Shelef G, Turzynski Berta (1979) Trickle irrigation using treated wastewaters. J Irrig Drain Eng ASCE 105: 175–187

Oster JD (1994) Irrigation with poor quality water. Agric Water Manage 25: 271–297

Pal D, Sen HS, Dash NB, Bandyopadhyay BK (1992) Moisture transport under cyclic trickle irrigation in different textured soils. J Agric Sci 118: 109–117

Parchomchuk P (1976) Temperature effects on emitter discharge rates. Trans ASAE 19: 690–692

Parkinson KJ (1985) Porometry. In: Marshal B, Woodward FI (eds) Instrumentation for environmental physiology. Soc For Exp Bot Semin Ser 22: 171–191 Camebridge Univ Press, Camebridge

Peacock WL, Rolston DE, Aljibury FK, Rauschkolb RS (1977) Evaluating drip, flood and sprinkler irrigation of wine grapes. Am J Enol Vitic 28: 193–19S

Or D, Groeneveld DP (1994) Stochastic estimation of plant available soil water in Owens valley under fluctuating water depths. J Hydrol 163: 63–64

Pelleg D, Lahav N, Goldberg SD (1974) Formation of blockages in drip irrigation systems, their prevention and removal. Proc 2nd Int Drip Irrig Congr ASAE Publ 105: 203–208

Penman HL (1948) Natural evaporation from open water, bare soil and grass Proc R Soc Lond 193: 120–146

Phene CJ (1986) Automation. In: Nakayama FS, Bucks DA (eds) Trickle irrigation for crop production. Design, operation and management. Elsevier, Amsterdam, pp 188–216

Phene CJ (1995) The sustainability and potential of subsurface drip irrigation. Proc Fifth Int Microirrig Congr Orlando, Florida ASAE Publ 4: 359–368

Phene CJ, Beale DW (1976) High-frequency irrigation for water-nutrient management in humid regions. Soil Sci Soc Am J 40: 430–436

Phene CJ, Beale DW (1979) Influence of twin-row spacing and nitrogen rates on high-frequency trickle-irrigated sweet corn. Soil Sci Soc Am J 43: 1216–1221

Phene CJ, Howell TA (1984) Soil sensor control of high frequency irrigation. Trans ASAE 27: 392–396

Phene CJ, Sanders DC (1976) High frequency irrigation and row spacing effects on yield and quality of potatoes. Agron J 68: 602–608

Phene CJ, Hoffman GJ, Rawlins SR (1971) Measuring soil matric potential in situ by sensing heat dissipation of a porous body Soil Sci Soc Am Proc 35: 27–33

Phene CJ, Howell TA, Beck RD, Sanders DC (1981) A traveling trickle irrigation system for row crops. "Irrigation, the hope and the promise". Annu Techn Conf Proc Irrigation Association, Arlington, Virginia, pp 66–81

Phene CJ, Blume MF, Hile MMS, Meek DW, Re JV (1983) Management of subsurface trickle irrigation systems, ASAE paper no 83-2598 ASAE, St. Joseph, Michigan

Phene CJ, Davis KR, Hutmacher RB, Bar-Yosef B, Meek DW, Misaki J (1991) Effect of high-frequency surface and subsurface drip irrigation on root distribution of sweet corn. Irrig Sci 12: 135–140

Phene CJ, Hoffman GJ, Rawlins SR (1971) Measuring soil matric potential in situ by sensing heat dissipation of a porous body Soil Sci Soc Am Proc 35: 27–33

Phene CJ, McCormick RL, Davis KR, Piero J, Meek DW (1989) A lysimeter feedback system for precise evapotranspiration measurements. Trans ASAE 32: 477–484

Philip JR (1968) Steady infiltration from buried point sources and spherical cavities. Water Resour Res 4: 1039–1047

Philip JR (1971) General theorem on steady-state infiltration from surface sources, with application to point and line sources. Soil Sci Soc Am Proc 35: 867–871

Philip JR (1984) Travel times from buried and surface infiltration point sources. Water Resour Res 20: 990–994

Philip JR (1986) Linearized unsteady multidimensional infiltration. Water Resour Res 22: 1717–1727

Philip JR (1991a) Upper bounds on evaporation losses from buried sources. Soil Sci Soc Am J 55: 1516–1520

Philip JR (1991b) Effects of root and subirrigation on evaporation and percolation losses. Soil Sci Soc Am J 55: 1520–1523

Philip JR (1992) What happens near a quasi-linear point source? Water Resour Res 28: 47–52

Pitts D, Bianchi M, Clark C (1995) Scheduling microirrigations for winegrapes using CIMIS. Proc Fifth Int Microirrig Congr Orlando, Florida ASAE Publ 4: 792–798

Plastro Gvat (1989) Filtration and water treatment manual. Plastro, Gvat, Israel

Plaut Z, Zieslin N (1977) The effect of canopy wetting on plant water status, CO_2 fixation, ion content and growth rate of 'Baccara' roses. Physiol Plant 39: 317-322

Plaut Z, Rom M, Meiri A (1985) Cotton response to subsurface trickle irrigation, Proc Third Int Drip/trickle Irrig Congr Fresno, California ASAE 2: 916-921

Pogue WR, Pooley SG (1985) Tensiometric meadsurement of soil water. Proc Third Int Drip/Trickle Irrig Congr Fresno, California ASAE 2: 761-766

Proebsting EL, Middleton JE, Roberts S (1977) Altered fruiting and growth characteristics of "Golden Delicious" apples associated with irrigation method. Hortscience 12(4): 349-350

Pruitt WO, Ferreres E, Henderson DW, Hagan RW (1981) Potential of drip irrigation in row crops for agricultural water conservation in California. Ann Rep Proj W-572, University of California Water Resources Center, Davis, California

Raats PAC (1972) Steady infiltration from sources at arbitrary depths. Soil Sci Soc Am Proc 36: 399-401

Radin JW, Reaves LL, Mauney JR, French OR (1992) Yield enhancement in cotton by frequent irrigation during fruiting. Agron J 84: 551-557

Rauschkolb RS, Rolston DF, Miller RJ, Carlton AB, Duran RG (1976) Phosphorus fertilization with drip irrigation. Soil Sci Soc Am J 40: 68-72

Ravelo CJ, Hiler EA, Howell TA (1977) Trickle and sprinkle irrigation of grain sorghum. Trans ASAE 20: 96-104

Ravina I, Paz E, Sofer A, Marcu A, Schischa A, Sagi G (1992) Control of emitter clogging in drip irrigation with reclaimed wastewater. Irrig Sci 13: 129-139

Ravina I, Paz E, Sagi G, Schisha A, Maren A, Yechieli Z, Sofer Z, Lev Y (1995) Performance evaluation of filters and emitters with secondary effluent. Proc Fifth Intern Microirrig Congr Orlando, Florida ASAE 4: 244-249

Rawitz E (1970) The dependence of growth rate and transpiration rate on plant and soil physical parameters under controlled conditions. Soil Sci 110: 172-182

Rawlins SL (1973) Principles of managing high frequency irrigations. Soil Sci Soc Am Proc 37: 626-629

Rawlins SL, Raats PAC (1975) Prospects for high-frequency irrigation. Science 188: 604-610

Reed AD, Meyer JL, Aljibury FK, Marsh AW (1977) Irrigation costs. Leaflet 2875 Division of Agricultural Sciencees, University of California, Davis, California p 10

Reeder BD, Newman JS, Worthington JW (1979) Effect of trickle irrigation on peach trees. Hortscience 14(1): 36-37

Reuther W (1944) Response of Deglet Noor date palms to irrigation on a deep sandy soil. 21st Ann Date Growers Inst, Coachella pp 16-19

Revol Ph BE, Clothier Lesaffre B, Vachaud G (1995) An approximate time-dependent solution for point-source infiltration. Proc Fifth Int Microirrig Congr, Orlando, Florida ASAE Publ. 4: 603-608

Risse LM, Chesness JL (1989) A simplified design procedure to determine the wetted radius for a trickle emitter, Trans ASAE 32(6): 1909-1914

Ritchie JJ (1972) Model for predicting evaporation from a row crop with incomplete cover, Water Resour Res 8: 1204-1213

Rochester EW (1995) Landscape irrigation design ASAE Publ 8

Rodney DR, Roth RL, Gardner BR (1977) Citrus responses to irrigation methods. Proc Int Soc Citric 1: 106-110

Russo D (1983) Leaching characteristics of a stony desert soil. Soil Sci Soc Am J 47: 431-436

Russo D (1984) Statistical analysis of crop yield - soil water relationships in heterogeneous soil under trickle irrigation. Soil Sci Soc Am J 48: 1402-1410

Russo D (1988) Determining soil hydraulic properties by parameter estimation: on the selection of a model for hydraulic properties. Water Resour Res 24: 453-459

Russo D (1993) A geostatistical approach to trickle irrigation design in heterogeneous soil. 1. Theory. Water Resour Res 16: 632-634

Rymon D, Fishelson G (1988) Economic analysis of cotton irrigation technologies. In: Rymon D (ed) Optimal yield management. Avebury, Aldershot, UK pp 221-235

Sagi G, Paz E, Ravina I, Schisha A, Marcu A, Yechieli Z (1995) Clogging of drip irrigation systems by colonial protozoa and sulfur bacteria. Proc Fifth Intern Microirr Congr Orlando, Florida ASAE Publ 4: 250-259

Sammis TW (1980) Comparison of sprinkle, trickle, subsurface and furrow irrigation methods for row crops. Agron J 72: 701-704

Scaife A, Bar-Yosef B (1995) Nutrient and fertilizer management in field grown vegetables. IPI-bulletin no. 13 International Potash Institute, Basel

Schischa A, Ravina I, Sagi G, Paz E, Yechieli Z, Alkon A, Schramm G, Sofer Z, Marcu A, Lev Y (1996) Clogging control in drip irrigation system used with reclaimed wastewater-"the platform trials". Proc 7th Int Conf on water and irrigation, Agritech, Tel Aviv pp 104-115

Schwartzman M, Zur B (1986) Emitter spacing and geometry of wetted soil volume, J Irr Drain Eng ASCE 112: 242-253

Searle SE (1954) Plant climate and irrigation. Chichester Press, Chichester, pp 112-134

Sefarim Y, Shmueli M (1975) The cost of water saved by using alternative irrigation systems. Hassadeh 55: 1182-1190 (in Hebrew with English summary)

Seginer I (1979) Irrigation uniformity related to the horizontal extent of the root zone. Irrig Sci 1: 89-96

Seifert WJ, Hiler EA, Howell TA (1975) Trickle irrigation with water of different salinity levels. Trans ASAE 18(1): 89-94

Sen HS, Paul D, Bandyopadhyay K, Dassh MB (1992) A simple numerical solution for two-dimensional moisture distribution under trickle irrigation. Soil Sci 154: 350-356

Shalhevet J, Mantell A, Bielorai H, Shimshi D (1981) Irrigation of field and orchard crops under semi-arid conditions, 2nd edn. International Irrigation Information Center (IIIC), Bet Dagan, Israel

Shalhevet J, Shimshi D, Meir T (1983) Potato irrigation requirements in a hot climate using sprinkler and drip methods. Agron J 75: 13-16

Shani U, Xue S, Gordin-Katz R, Warrick AW (1996) Soil-limiting flow from subsurface emitters. I. Pressure measurements. J Irrig Drain Eng ASCE 122: 291-300

Shearer MN, Martin LW, Lombard PD (1975) Drip irrigation research in Oregon-a progress report. Proc Second Int Drip Irrig Congr ASAE Publ 105: 39-43

Shmueli M (1975) Drip irrigation of vegetables with saline water. Hortscience 10: 506-509

Shoji K (1977) Drip irrigation. Sci Am 237: 62-68

Shouse P, Jury WA, Stolzy LH, Dasberg S (1982) Field measurement and modeling of cowpea water use and yield under stressed and well-watered growth conditions. Hilgardia 50: 6

Shuval HI (1977) Health considerations in water renovation and reuse. In: Shuval HI (ed) Water renovation and reuse. Academic Press, New York, pp 33-73

Simunek J, Vogel T, van Genuchten MT (1993) SWMS 2D: simulating water flow and solute transport in two-dimensional variably saturated media. US Salinity Lab, Riverside, CA

Smajstrla AG (1993) Microirrigation for citrus production in Florida, Hortscience 28: 295-298

Smith SW, Walker WR (1975) Annotated bibliography on trickle irrigation. Colorado State University information series no 6 Environmental Resources Center, Fort Collins, Colorado, p 65

Solomon K (1977) Performance comparison of different emitter types. Proc 7th Int Agric Plastics Congr, San Diego, pp 97-102

Solomon K (1979) Manufacturing variation of trickle emitters. Trans ASAE 22(5): 1034-1043

Solomon K, Keller J (1978) Trickle irrigation uniformity and efficiency. J Irrig Drain Eng ASCE 104: 293-306

Spaans EJA, Baker J (1992) Calibration of Watermark soil moisture sensors for soil matric potential and temperature. Plant Soil 143: 213-217

Stevenson DS (1981) Responses of 6-year-old Diamond grapevines to the change from sprinkler to trickle irrigation and to the time and method of applying nitrogen. Can J Soil Sci 61: 571–575

Stibbe E (1986) A comparison between the gross energy requirements of different irrigation systems in Israel. Irrig Sci 7: 213–224

Taghavi SA, Marino MA, Rolston DE (1984) Infiltration from a trickle irrigation source. J Irrig Drain Eng ASCE 110: 331–341

Tajirshy MA, Hills DJ, Tchobanoglous G (1994) Pretreatment of secondary effluent for drip irrigation. J Irrig Drain Eng ASCE 120: 716–731

Tanner CB, Jury WA (1976) Estimating evaporation and transpiration from a row crop during incomplete cover. Agron J 68: 239–243

Taylor SA (1965) Managing irrigation water on the farm. Trans ASAE 8: 433–436

Taylor SA, Ashcroft AB (1972) Soil edaphology. Freeman, San Francisco

Tollefson S (1985) Subsurface drip irrigation of cotton and small grains. Proc Third Int Drip/Trickle Irrig Congr Fresno, California ASAE 2: 887–895

Tomer E, Shooker S, Ripa M (1995) Response of avocado trees to different irrigation regimes. Alon Hanotea 49: 445–451 (in Hebrew, with English summary)

Topp GC, Davis JL, Annan AP (1980) Electromagnetic determination of soil water content; measurements in co-axial transmission lines. Water Resour Res 16: 574–582

Torrecilas A, Ruiz-Sanchez MC, Leon A, DelAmor F (1989) The response of young almond trees to different drip-irrigated conditions. Development and yield. J Am Soc Hortic Sci 114: 25–29

Tscheschke P, Alfaro JF, Keller J, Hanks J (1974) Trickle irrigation soil water potential as influenced by management of highly saline water. Soil Sci 117: 226–231

Turner NC (1987) The use of the pressure chamber in the measurement of plant water status. Proc Int Conf on Measurement of Soil and Plant Water Status Utah State University, Logan, Utah 2: 13–14

Van Bavel MG (1995) Advances in microirrigation control by sap flow monitoring systems. Proc Fifth Int Microirr Congr Orlando, Florida ASAE Publ 4: 234–238

Van Bavel CHM, Newman JE, Hilgeman RH (1967) Climate and estimated water use by an orange orchard. Agric Meteorol 4: 27–32

Van Genuchten MT (1980) A closed-form equation for predicting the hydraulic conductivity of unsaturated soils. Soil Sci Soc Am J 44: 892–898

Van Genuchten MTh, Wierenga PF (1986) Solute dispersion coefficients and retardation factors. In: Klute A (ed) Methods of Soil Analysis Part 1. Monograph no 9 ASA, Madison, Wisconsin, pp 1025–1054

Van Goor RJ, De Jager A, Voogt W (1988) Nutrient uptake by some horticultural crops during the growing period. Proc Seventh Int Cong. on Soilless Cult Wageningen, Netherlands pp 163–175

Walker WR (1979) Explicit sprinkler uniformity efficiency model. J Irrig Drain Eng ASCE 105: 129–136

Walker WR, Smith SW, Geohring L (1976) Evapotranspiration potential under trickle irrigation. Paper no 76-200 g. ASAE, St Joseph, Michigan

Wallach R (1990) Effective irrigation uniformity as related to root zone depth. Irrig Sci 11: 15–21

Warrick AW (1974) Time dependent linearized infiltration I Point sources. Soil Sci Soc Am Proc 38: 383–386

Warrick AW (1983) Interrelationships of irrigation uniformity terms. J Irrig Drain Eng 109: 317–332

Warrick AW (1985) Point and line infiltration – calculation of the wetted soil surface. Soil Sci Soc Am J 49: 1581–1583

Warrick AW (1986) Design principles. In: Nakayama FS, Bucks DA (eds) Trickle irrigation for crop production. Design, operation and management. Elsevier, Amsterdam, pp 93–116

Warrick AW, Lomen DO, Amoozegard-Fard A (1980) Linearized moisture flow with root extraction for three-dimensional, steady conditions. Soil Sci Soc Am J 44: 911–914

Waterfield (1973) Trickle irrigation in the United Kingdom. In: FAO irrigation and drainage paper vol 14. FAO, Rome, pp 147–152

Watters GZ, Keller J (1978) Trickle irrigation tubing hydraulics. ASAE paper no 78-2015. St Joseph, Michigan

Wierenga PJ (1977) Influence of trickle and surface irrigation on return flow quality. Robert S Kerr Environmental Research Laboratory, Office of Research and Development, EPA, Ada, Oklahoma

Willoughby P, Cockroft B (1974) Changes in foot pattern of peach trees under trickle irrigation. Proc 2nd Int Drip Irrig Congr ASAE Publ 105: 439–442

Wooding RA (1968) Steady infiltration from a circular pond. Water Resour Res 4: 1259–1273

Wu IP (1975) Design of drip irrigation main lines. J Irrig Drain Eng ASCE 101: 265–278

Wu IP (1985) A simple optimal microirrigation scheduling. Proc Fifth Int Microirr Congr Orlando, Florida ASAE Publ 4: 781–786

Wu I-P, Gitlin HM (1973) Hydraulics and uniformity for drip irrigation. J Irrig Drain Eng ASCE 99: 157–168

Wu I-P, Gitlin HM (1975) Irrigation efficiencies of surface, sprinkler and drip irrigation Proc 2nd World Congr on Water Resource Water for human needs, New Delhi, India pp 191–199

Wu I-P, Gitlin HM (1977) Drip irrigation system design in metric units. Miscellaneous Publications of Cooperative Extension Service, no 144 University of Hawaii, 19 pp

Wu I-P, Gitlin HM (1979) Drip irrigation design on nonuniform slopes. J Irrig Drain Engng, ASCE 105: 289–303

Wu IP, Howell TA, Hiler EA (1979) Hydraulic design of drip irrigation systems. Hawaii Agric Exp Stn Tech Bull 105, Honolulu

Wu IP, Gitlin HM, Solomon KH, Saruwahari CA (1986) System design. In: Nakayama FS, Bucks DA (eds) Trickle irrigation for crop production. Elsevier, Amsterdam pp 53–92

Yagev E (1977) Drip irrigation in citrus orchards. Proc Int Soc Citric 1: 110–113

Yaron B, Shalhevet J, Shimshi D (1973) Patterns of salt distribution under trickle irrigation. In: Hadas A (ed) Physical aspects of soil water and salts in ecosystems. Springer, Berlin Heidelberg New York pp 389–394

Yarwood CE (1978) Water and the infection process. In: Kozlowsky TT (ed) Water deficits and plant growth, vol 5 Academic Press, New York, pp 141–156

Zohar Y (1971) Observations and field experiments testing trickle irrigation in vegetable crops. In: Drip (trickle) and automated irrigation in Israel, vol 1. Water Commissioner's Office, Ministry of Agriculture, Tel Aviv, pp 117–131

Zoldoske DF, Genito S, Torgensen GS (1995) Surface drip irrigation on turfgrass, a university experience. Proc Fifth Intern Microir Congr Orlando, Florida ASAE publ 4: 300–302

Zur B (1996) Wetted soil volume as a design objective in trickle irrigation. Irrig Sci 16: 101–105

Zur B, Tal S (1981) Emitter discharge sensitivity to pressure and temperature. J Irrig Drain Eng ASCE 107: 1–9

Appendix: Case Studies

1
Apple Orchard

An example of a drip irrigation design in an orchard will be given here, utilizing the principles given in Chapter 4. The information available is as follows:

The proposed apple orchard is of irregular shape, has an area of approximately 5 ha and a varying slope in the north-south direction, as indicated in Fig. A.1. There is no slope in the east-west direction. A water outlet providing $20\,m^3/h$ of high quality well water at a pressure of 20 m is present at point A. The pressure loss caused by control valves and filters is estimated to be 5 m. The proposed planting distances are 4×8 m, with row direction north – south. The soil is a silt loam, with the following characteristics: saturated hydraulic conductivity Ks = $0.2\,m^3/h$, $\alpha = 0.014\,cm^{-1}$ (see Table 3.1). It is proposed to use in-line long-path emitters, which are available with a discharge rate of 2, 3, 4 or 8 l/h at any desired distance on the lateral. The available lateral sizes are 8, 12, 16, 20, 24 and 30 mm diameter. For main lines the following sizes are available: 30-, 36-, 42-, 40- and 60-mm diameter plastic pipes. The design should allow a maximum pressure variation of 20%, which is a common design criterion.

Determination of Distances Between Emitters

It may be assumed that the most economical way to design a drip system in an orchard is to provide one lateral per tree row, resulting in a wetted strip. The discharge and spacing between emitters should result in an overlap of saturated radii of water entry ponds. The first questions which will arise therefore are which of the available emitters to use and at what distance apart the emitters should be placed. The procedure given in Section 4.2.2 can provide us with some guidelines. Using the information on the soil hydraulic properties and the available emitter discharge rates, the radius of the saturated zone is calculated using Eq. (4.5) as follows:

Emitter discharge (l/h)	2	3	4	8
Saturated radius (cm)	26.9	37.3	46.4	76.2
Emitter spacing (cm)	54.0	74.5	92.7	152.0

It appears that 4 l/h emitters spaced 1 m apart will provide adequate soil wetting along the laterals, while with 8 l/h emitters the ponded area between the tree rows may be too large. Therefore we chose the 4 l/h emitters spaced 1 m apart. It has to be borne in mind that this calculation procedure can provide only a general guideline because of the many simplifications on which it is based.

Fig. A.1. A schematic map of the apple orchard for case study 1

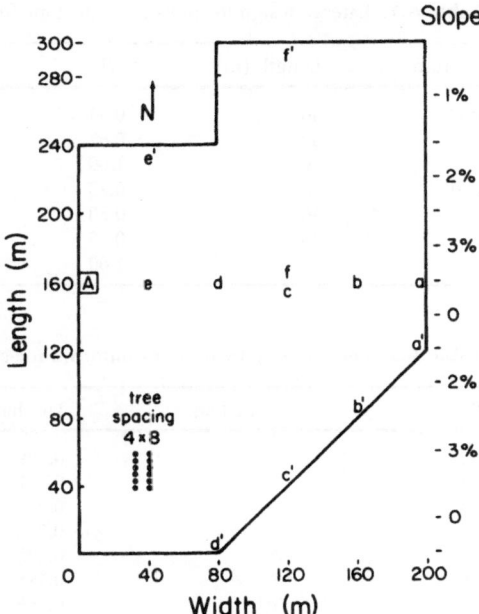

Lateral Design

There are several possibilities for lateral design, based on the location of the main line (see Fig. A.1). Since the water source is located at point A, the main line can be put in a west-east direction starting at point A, approximately in the middle of the field. This will result in some of the laterals leading upslope in the northern half of the field. Moving the water source to the northern edge of the field will cause all the laterals to go downslope. However, they will be rather long (240-m length) and an additional length of main line is needed. Both options will be shown.

a. Main line in middle of field.
Laterals will be designed at several locations, using various procedures.

a-a′ This 40-m long lateral has a total discharge of 160 l/h = 0.044 l/s. L/H = 2.67, and the slope is 0. Using Fig. 4.1, these data give a value of $\Delta H/L = 6\%$. Matching this value with Q 0.044 l/s in the nomograph of Fig. 4.2 will give a tube diameter of 8 mm.

b-b′ A similar procedure, assuming uniform slope of 1%, will give a 12-mm tube for this 80-m-long lateral.

c-c′ For this 120-m-long lateral, the procedure for non-uniform slopes is applied (see Table A.1). Points 1/L, Hi/L are plotted in quadrant I of Fig. 4.3. Assuming a tube diameter of 12 mm at Q = 0.133 l/s gives $\Delta H/L = 0.05$ in Fig. 4.2. $\Delta H = 6$, $\Delta H/H = 0.05$, L/H = 8. These data, plotted in Fig. 4.3, result in an acceptable design with less than 20% pressure variation.

d-d′ The data for this lateral are also given in Table A.1. Taking a tube diameter of 16 mm and following the same procedure as outlined above, this leads to an

Table A.1. Lateral design for non-uniform slope for the apple orchard

Section	Length (m)	1/L	Slope (%)	H_1 (m)	H_1/L
c-c'	40	0.33	0	0	0.0
	40	0.67	2	0.8	0.0066
	40	1.00	3	2.0	0.016
d-d'	40	0.25	0	0	0
	40	0.50	2	0.8	0.095
	40	0.75	3	2.0	0.0125
	40	1.00	0	2.0	0.0125

Table A.2. Lateral design with non-uniform tube size for the orchard, section f'-f-c-c'

Section	Slope (%)	Discharge (L/s)	Tube diameter (mm)
1	1	0.289	30
2	1	0.267	30
3	1	0.244	24
4	2	0.222	24
5	2	0.200	24
6	3	0.178	20
7	3	0.155	20
8	0	0.133	20
9	0	0.111	20
10	2	0.089	16
11	2	0.067	16
12	3	0.044	12
13	3	0.022	8

acceptable design with less than 10% pressure variation. When a 12-mm tube is tried, however, a friction drop of 9% results, which would leave essentially no pressure at the end of the 160-m lateral.

e-e' For this upslope lateral, Fig. 4.1b will be used. A uniform slope of 2.5% will be assumed, L/H = 5.3. These data plotted in Fig. 4-1b give $\Delta H/L = 1.5$. This value with Q = 0.09 l/s show, using Fig. 4-2, that 12 mm is too small a diameter, therefore 16 mm is the acceptable diameter for this lateral.

f-f' A similar procedure as for e-e' leads to 20 mm for this lateral.

b. Main line at edge of field

Two options will be shown for planning the 260-m-long lateral f'-f–c-c', for the case of the main line placed at the northern edge of the field. Applying the principle of varying slope, as in the case of c-c' in the previous section, taking Q = 0.29 l/s and a tube diameter of 20 mm leads to an acceptable design. A tube line of 16 mm, however, will fall outside the acceptable region (the reader is required to go through the detailed procedure in order to arrive at this conclusion). It seems more appropriate to use the design method for varying tube sizes in this case. The length of the lateral is divided into sections of 20 m, and for each section the discharge is calculated (see Table A.2). Now the nomograph of Fig. 4.4 is used, matching the slope for each section with its discharge. The results given in Table A.2 show a tube size varying from 8 mm diameter for the last 20 m of lateral to 30 mm for the first 40 m of lateral. It should be noted that this simplified method can only be used in a downslope situation.

Comparing the results of this design with the case of the main line in the center of the field, it is clear that the latter design is preferable. Not only are the lateral tube sizes smaller, but also less main line is needed. No precise cost calculations are needed in order to arrive at the conclusion that the main line should be placed at the center of the field.

Main Line Design

In order to design the pipe size for the main line, some assumptions are needed for the irrigation schedule. If all drippers are in operation, the total discharge at point A would be approximately 6.67 l/s. Obviously, the orchard can be irrigated in sections. If the peak irrigation demand is assumed to be 5 mm/day (equivalent to 160 l/tree) and taking into account that each tree is irrigated by four drippers of 4 l/h, this means that with daily irrigation the maximum irrigation time is about 10 h. The conclusion from this is that in order to meet peak demand, the orchard can only be divided into two sections. The maximum discharge at point A is therefore 3.33 l/s. The other necessary assumption is about the energy slope. Since the pressure head at point A is 15 m and assuming that the minimum pressure at the head of each lateral is 10 m (no slope in the field in the E-W direction), the energy slope over 200 m of main line will be 5/200 = 2.5%. Dividing the main line into five sections, the maximum discharge starting at point A would be 3.3, 2.67, 2.00, 1.22 and 0.55 l/s. The nomograph of Fig. 4.4 shows that these discharges will correspond to pipe diameters of 60, 56, 50, 42 and 30 mm, respectively, taking an energy slope of 2.5%.

2
Cotton

This case study deals with the irrigation design of 100 ha cotton on a clay loam soil. Cotton is grown in 1.92-m spaced beds with two rows per bed. The maximum topographical difference is 6 m, resulting in slopes of 0–2%. The project was designed by Naan Irrigation Systems, with the help of the software of WCADI (Weizman Industies LTD, Israel). The basic data of the design will be presented and will be evaluated according to the principles given in Chapter 4.

Water Supply

The area will be irrigated by 16 wells with a discharge of 3–12 l/s. The total discharge of all the wells is 250–300 m³/h. The assumed peak rate of evapotranspiration is 5 mm/day, this is equivalent to 500 m³/day for 100 ha (200 m³/h). The discharge of the wells is therefore adequate to supply the irrigation demand. The wells will be connected into an operational reservoir, from which the water will be pumped into the irrigation main lines. The size of the reservoir should enable the supply of at least 8 h of irrigation and was chosen to be 2500 m³. The pump station will include three 150 m³/h pumps. Two pumps are designed for continuous operation and the third pump will be a reserve. A computer, activated by a control float in the reservoir, will control all the pumps, including those on the wells.

Filtration

No data were available on water quality at the time of system design, except for the fact that water will be supplied by wells. A battery of six gravel filters was proposed for a maximum flow rate of $250\,m^3/h$, with automatic flushing. At each submain irrigating 2 ha a check screen filter will be installed requiring periodic manual cleaning. The laterals should be flushed manually once a week.

Choice of Emitters

The choice of the emitters in this design was based on experience with cotton irrigation and on the availability of the equipment. Naan-Paz 35 was chosen with the following characteristics: It is an integral molded labyrinth-type emitter with an exponent of 0.5 and CV < 0.03. The wall thickness is 0.9 mm, the inner diameter is 15.4 mm. The emitter flow rate at a pressure of 15 m is 1.9 l/h and at 20-m pressure it is 2.4 l/h. The minimum lifetime of the laterals, with annual retrieval and installation, is 8 years. The maximum lateral length recommended by the manufacturer in a flat area for 10% flow variation at an emitter spacing of 50 cm is 118 m.

In trying to evaluate the choice of the emitters and their proposed distances (50 × 192 cm) according to the principles given in Chapter 4 we are confronted with the lack of data on the soil characteristics; a rather common situation in irrigation design. We propose an inverse procedure based on Eq. (4.4) in order to calculate the soil hydraulic conductivity K_s according to

$$K_s = \left(q/0.83\,V_w^*\right)\left(z^2/d^2\right),$$

Where q = the emitter discharge, V_w^* = the wetted soil volume given by $V_w^* = 13.5$ PI.DL = $2592\,cm^2$ (PI = preferred irrigation interval, 1 day in our case and DL = the distance between the laterals, 192 cm), d = distance between emitters (50 cm) and z = depth of wetting. Assuming a hemispheric wetted volume (d = z) we arrive at a value of $K_s = 2000\,[cm^3/h]/0.83 \times 2592\,[cm^2] = 1\,cm/h$ ($2.8 \times 10^{-4}\,cm/s$) which is an approximate value for a medium-textured soil (see Table 3.1). We may conclude that the choice of emitters and their distances is reasonable under the given circumstances. If, on the other hand, the calculated soil hydraulic conductivity was much higher than the estimated value for this soil type, then emitter discharge should have been adjusted accordingly (i.e. reduced).

Lateral Design

The area is divided into 8 units of approximately 12 ha each to be irrigated in series (see Fig A.2 for an overview of the project). Each unit is subdivided into 2-ha blocks, served by 75-mm submains and laterals of 100–110 m length (see Fig A.3 for a detailed view). The laterals are partly upslope and partly downslope, the maximum slope is 2%. We apply the procedure of Section 4.3 using Figs 4.1 and 4.2 in order to evaluate the tube diameter of the chosen laterals. At a given inlet pressure of 20 m over a length of 100 m and a total discharge of 480 l/h (0.13 l/s), the lateral tube diameter should be 12 mm for 2% downslope and 14 mm for 2% upslope. (The reader should verify the procedure). The chosen lateral with 15.4 mm inner diameter is therefore well within the desired pressure variation of

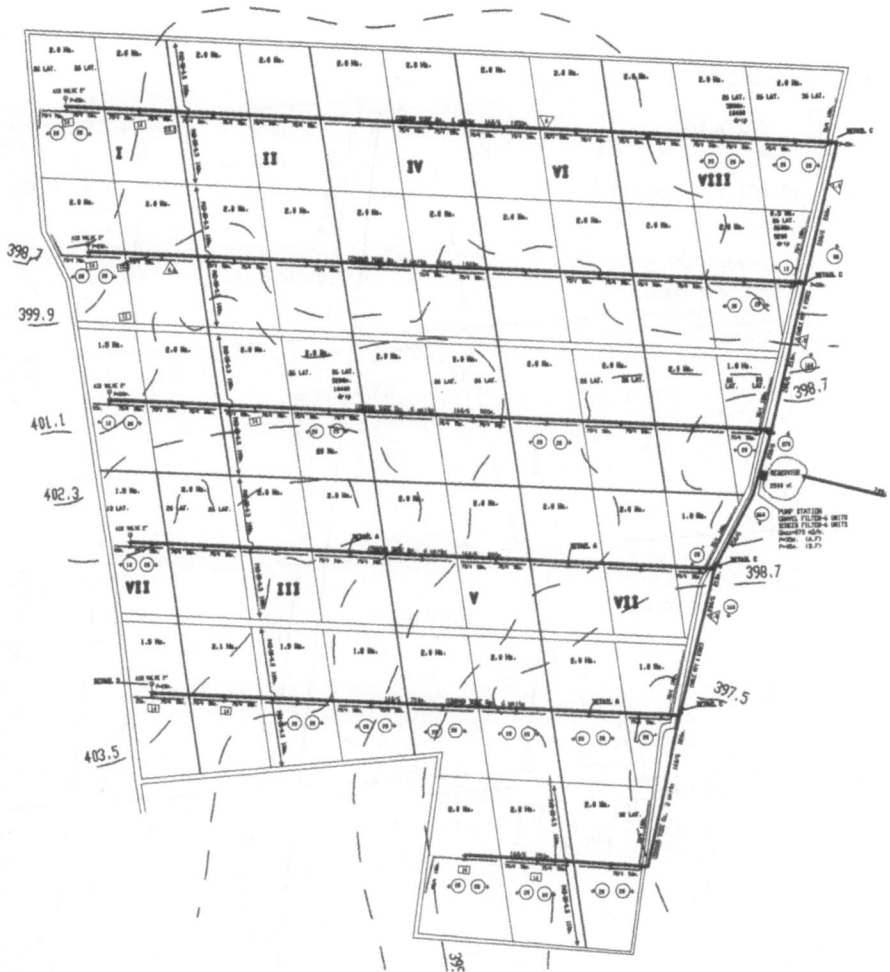

Fig. A.2. An overview of the cotton project for case study 2

20% resulting in an outflow variation of 10% in accordance with the specifications given by the manufacturer.

Submains of 75-mm diameter are designed for an average flow rate of 20 m³/h (serving two sets of 26 laterals). According to Fig. 4.3 the head loss should be 2 m/100 m or 1 m/50 m submain. Acording to the computer design, the head loss is 0.6 m, which is somewhat lower.

Main Line Design

Ten submains of 75-mm diameter are served by a main line of 160-mm diameter. For a flow of 80 m³/h (four 2-ha blocks to be irrigated simultaneously) this will result in

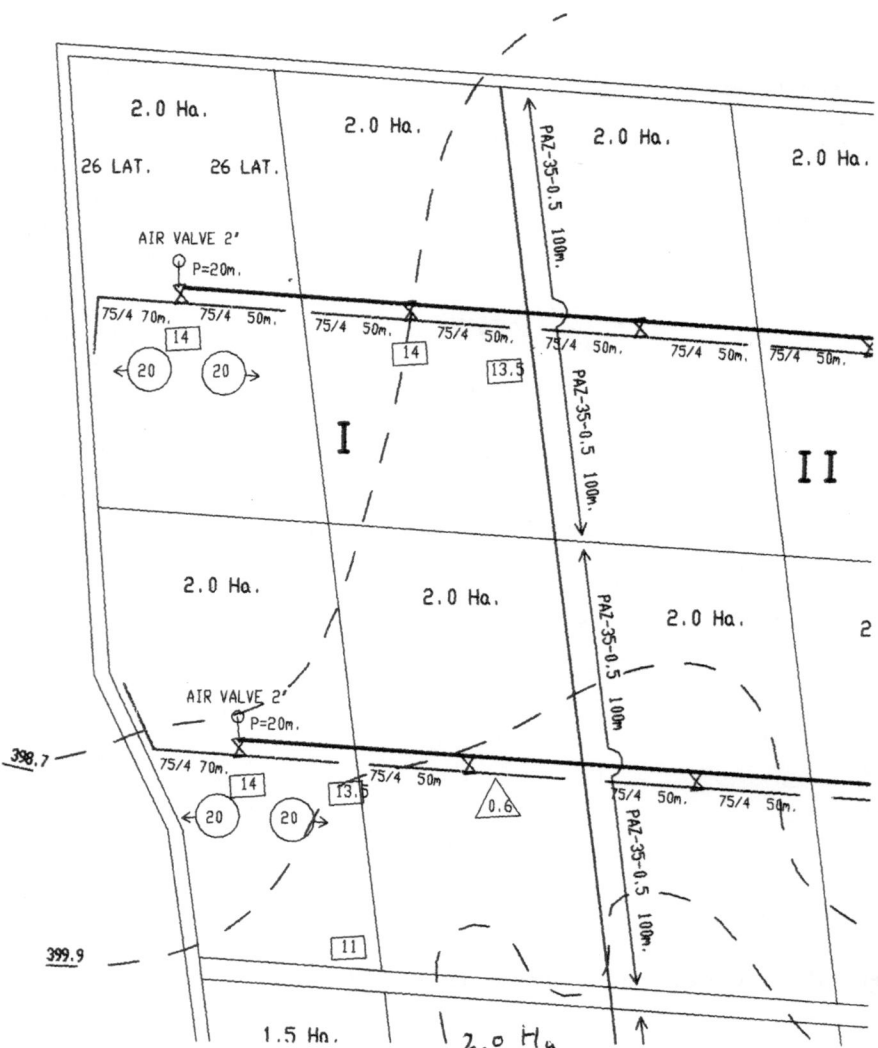

Fig. A.3. A detailed view of unit I (11.5 ha) of the cotton project for case study 2

a head loss of 0.6 m/100 m or 6 m/km, which is according to the computer design. The main line of 200-mm diameter, serving an area of 10 ha to be irrigated simultaneously at a flow of 160 m³/h, should give a head loss of 1 m/100 m according to Fig. 4.3. The total head loss along the main line irrigating a unit of 12.5 ha at 250 m³/h is 2.5 m according to the computer design and this includes a section of 100 m of 250-mm diameter pipe. It is clear that in this more complicated case, the computer–aided design will give an accurate and optimal solution for the selection of laterals, mains and submains, based on operational constrains.

Management

The irrigation will be controlled by a computer, activated by a central water meter with electronic pulses, enabling serial irrigation of the eight units at a rate of 230–270 m^3/h each at a pressure of 15 m for a duration of approximately 2.5 h daily each, resulting in 20 h/day of irrigation at the peak rate of 5 mm/day. The irrigation time can be shortened to 16 h/day by a larger operating pressure (20 m) resulting in an increased emitter flow rate of 2.4 l/h. This will allow more flexibility in the management, allowing time for spraying and maintenance of the system.

Fertigation is controlled by electric fertilizer pumps activated by the irrigation computer.

Subject Index